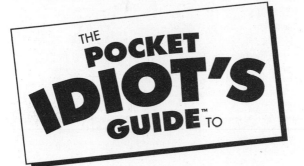

Your Carbon Footprint

by Nancy S. Grant

ALPHA

A member of Penguin Group (USA) Inc.

This book is for GCG and JAG.

ALPHA BOOKS

Published by the Penguin Group

Penguin Group (USA) Inc., 375 Hudson Street, New York, New York 10014, USA

Penguin Group (Canada), 90 Eglinton Avenue East, Suite 700, Toronto, Ontario M4P 2Y3, Canada (a division of Pearson Penguin Canada Inc.)

Penguin Books Ltd., 80 Strand, London WC2R 0RL, England

Penguin Ireland, 25 St. Stephen's Green, Dublin 2, Ireland (a division of Penguin Books Ltd.)

Penguin Group (Australia), 250 Camberwell Road, Camberwell, Victoria 3124, Australia (a division of Pearson Australia Group Pty. Ltd.)

Penguin Books India Pvt. Ltd., 11 Community Centre, Panchsheel Park, New Delhi—110 017, India

Penguin Group (NZ), 67 Apollo Drive, Rosedale, North Shore, Auckland 1311, New Zealand (a division of Pearson New Zealand Ltd.)

Penguin Books (South Africa) (Pty.) Ltd., 24 Sturdee Avenue, Rosebank, Johannesburg 2196, South Africa

Penguin Books Ltd., Registered Offices: 80 Strand, London WC2R 0RL, England

Copyright © 2008 by Nancy S. Grant

International Standard Book Number: 978-1-59257-774-3
Library of Congress Catalog Card Number: 2007941486

10 09 08   8 7 6 5 4 3 2 1

Interpretation of the printing code: The rightmost number of the first series of numbers is the year of the book's printing; the rightmost number of the second series of numbers is the number of the book's printing. For example, a printing code of 08-1 shows that the first printing occurred in 2008.

Printed in the United States of America

Contents

Introduction

Understanding your carbon footprint is a lot different than knowing your shoe size—you can't *see* your carbon footprint, but it's every bit as real as a size 8½ cross-trainer or pair of old gray athletic socks.

Although your carbon footprint is invisible, you can get an idea of its size. You'll need to know a bit of physics, review a dash of chemistry, and understand a sprinkle of math to see all the things that contribute to your carbon footprint. What you *won't* need is a fancy multi-key calculator or an old-fashioned slide rule. This book gives you handy reference points instead.

You'll also need to learn about many different forms of energy—including how they're produced and used—to see how the choices you make every day of the year influence the size of your carbon footprint. You'll see the impact your decisions have on the size of your carbon footprint.

That's the great thing about a carbon footprint—once you know how to measure it, you're not stuck with its size forever, and it's a lot easier than you might think to change it. This book gives you the tools to evaluate the consequences of all sorts of activities and helps you see how even simple decisions can make a big difference.

I spend a great deal of time writing about technical stuff; interviewing engineers and scientists, technicians and computer experts; and then explaining how they do their work to curious folks just like you. These days, lots of people want to know more about the science behind the things we take for granted in the twenty-first century. You already know a lot more science than you might give yourself credit for, and I'll explain things in a way that will build on your knowledge.

There's no mystery when dealing with making wise energy choices as long as you keep one principal in mind: every form of energy has its plusses and minuses. Choices involve money, convenience, short- and long-term environmental impacts, and practical problems to be solved.

This book is not about telling you what you should never, ever do, or making you feel guilty about a certain activity. Spending the weekend puttering around the lake on a gasoline-powered houseboat with friends is a great way to relax. So is sailing a schooner out on the ocean or paddling a canoe down a secluded creek, and one is not inherently better or worse than the other from an energy standpoint. But each one does involve a whole chain of energy choices and opportunities. My goal is to help you see that you really can make a difference, sometimes with tiny steps and sometimes with big strides.

As you'll soon see, one of the most important concepts to put into use right now to change the size of your carbon footprint is to follow the advice of an old slogan: reduce, recycle, reuse. As we explore your carbon footprint, I'll give you some fresh ideas on how to do each one.

Extras

I'm glad you're paying closer attention to how your choices and your use of human technology can affect the natural world we live in. Throughout this book, you'll find helpful info in boxes like these to help you understand and remember useful details.

def•i•ni•tion

Here you'll find definitions of important terms.

Carbon Impact

Check these boxes for insider knowledge and bits of easy-to-use advice.

Carbon Caution

These boxes point out situations in which you, the reader, must beware of dangers or scams or wastes of time and effort.

Carbon Extras

Here you'll find some odd yet interesting tidbits, info that could contradict commonly held ideas, or info to give you a deeper understanding of an issue.

Acknowledgments

First, thanks to Paul Wesslund, editor at *Kentucky Living* magazine, for offering me, years ago, the ongoing assignment as "Future of Electricity" columnist. He said I'd meet a lot of interesting people on my way to becoming an energy expert, and I have. I continue to appreciate Paul's confidence in my work.

Second, thanks to my family for tolerating the working methods of a freelance nonfiction writer. My husband, Glen, a retired construction electrician, accepted a long time ago that putting words and ideas together on a computer screen is every bit as complicated as wiring a building—and there's never a blueprint or schematic to check! Our daughter, Jennifer, an engineering student, has been an invaluable source for cellphone technology updates and practical computer advice. And our long-whiskered family cat has shown remarkable tolerance for being moved from one office chair to another when what I need to do to meet a deadline conflicts with his chosen napping spot.

Thanks, too, to longtime friend Julia Martin, high school science teacher extraordinaire. Decades of grading papers have shown her there's no concept in chemistry, physics, or earth science that can't be mangled. But she has the patience to explain it all over again one more time and the diligence to track down obscure facts. It's great for a science writer to have such a willing sounding board just an e-mail away.

And I want to thank the hundreds of experts who've talked with me over the years about their energy projects. Whether it's an electric utility company lineman explaining how GPS technology makes it easier to pinpoint the location of an outage and restore service promptly, or a university scientist setting up controlled experiments to evaluate a new application of technology, or a farmer who has installed heat-recapturing systems in a dairy barn—everyone I talk with is eager to share their ideas. Wise energy use is the most widespread, most talked-about, most important topic I've ever covered as a writer.

I thank you, the reader, too, because *your* interest in energy and the environment will make a difference for all of us.

Special Thanks to the Technical Reviewer

The Pocket Idiot's Guide to Your Carbon Footprint was reviewed by an outstanding expert who double-checked the accuracy of what I've included, to help make certain that this book gives you the facts you need. Special thanks are extended to Amanda Abnee Gumbert.

Trademarks

All terms mentioned in this book that are known to be or are suspected of being trademarks or service marks have been appropriately capitalized. Penguin Group (USA) cannot attest to the accuracy of this information. Use of a term in this book should not be regarded as affecting the validity of any trademark or service mark.

Carbon Footprint Basics

In This Chapter

- Understanding global climate change
- The natural carbon cycle
- Identifying greenhouse gases
- How humans affect the carbon cycle
- Measuring carbon footprints

You've probably seen and heard quite a few dire predictions about how people are causing global climate change. And you're just as likely to have run across opponents of the whole idea that humans need to worry about the consequences of their actions. How do you sort out the claims and counterclaims? How can what you do as an individual—just one person among more than 6 billion other human beings—have a measurable effect on anything as big as an entire planet?

This book helps you sort through the buzzwords and catchphrases that keep cropping up in discussions about the impact of humans on the natural world. I'll show you the facts and figures behind the broad concepts, and give you the tools you need to evaluate many aspects of energy use and your impact on the global environment. Then you can make well-informed decisions in everyday life and decide for yourself if your carbon footprint's size needs to be adjusted to something a little smaller.

Is Climate Change Real?

It's true that global climate change has occurred naturally many times in Earth's history, for reasons scientists are still investigating. During some periods, Earth has been much hotter or much colder than it is today.

Today the phrase "global climate change" commonly refers to one specific time period— now—and one particular cause—humans. Global climate change is shorthand for the idea that the actions of people all over the planet have reached such a critical point that we are interfering with the pace and direction of natural climate fluctuations. During a major period of global climate change about 14,000 years ago, glaciers that had extended south from Canada to the Ohio River Valley, and other glaciers in other parts of the Northern Hemisphere, rapidly melted and retreated. Local weather patterns on each continent changed.

How can you compare one action to another? Is there a way to measure your environmental impact that's as simple as reading the changing level of water in a rain gauge? Unfortunately, no one has yet invented a cheap, reliable, easy-to-read, all-purpose environmental impact gauge. But you can use your imagination to visualize how your actions affect the natural world.

Carbon Extras

Weather is local and varies over short periods of time. For example, the weather on Tuesday in Cincinnati might be rainy and cool, but the weather in Louisville might be sunny. Climate covers large geographic areas and tends to have repeating patterns. The climate in the Ohio River Valley is temperate, with four distinct seasons of cold winters with occasional periods of snow, mild springs with ample rain, hot and humid summers, and fairly dry autumns.

Your *carbon footprint* is invisible—you can't look down at the kitchen floor and see it the way you might notice snow from your boots after a walk on a wintry afternoon. But this imaginary symbol represents the focus of intense interest today. Understanding the size of your carbon footprint and the various components that go into it is becoming extremely important. In fact, the size of your carbon footprint may become one of the most important numbers in your life in the twenty-first century.

def•i•ni•tion

Carbon footprint sums up the impact human activities have on the environment in terms of the amount of greenhouse gases produced. A carbon footprint includes many things, but for measuring convenience, all factors are converted to units of carbon dioxide (CO_2).

As the number of people convinced that human actions affect the global climate increases, they're much more likely to seek action and change. New rules and regulations—at the local, national, and international levels—could affect the choices you have as an individual. Restrictions on all kinds of activities could affect what you pay for goods and services, or cause certain options and items to become scarce or nonexistent. Inventions and new ways to apply technology could change many aspects of daily life. The search for new and better ways to use energy could make a huge difference in the size of your carbon footprint.

Our Original Carbon Planet

Let's get started with a look at the *carbon cycle*. Carbon appears in many forms throughout our planet. Carbon is a part of every living thing, even dead things that are decomposing and things that have been dead for so long that just their fossil traces remain.

def•i•ni•tion

> The **carbon cycle** refers to the way carbon in its many forms moves around through various parts of the natural world.

Carbon is an *element*. Carbon *molecules* can be alone, or they can move around to combine with other elements. In one form or another, carbon is under our feet in the soil and rocks, around us in the plants and animals and people we live among, in the foods we eat, and in the air we breathe. Carbon is one of the building blocks of life.

def•i•ni•tion

> An **element** is a substance that cannot be reduced to a simpler substance by normal chemical means. A **molecule** is the smallest physical bit of an element that can exist and still have all the characteristics of that element. All molecules of an element have the same composition, but the individual parts may be aligned differently.

Carbon is the sixth most abundant element in the universe. On Earth, pure carbon occurs in two natural forms, as a diamond and as graphite.

Both diamonds and graphite are found in metamorphic rocks, areas that have undergone changes in pressure and temperature over very long periods of time. Oil and coal are not pure carbon; each is a mixture of many other elements. Oil, coal, and other fossil fuels are found in sedimentary rock formations. Before people came along, carbon molecules moved around in a completely natural cycle through both brief and incredibly long time periods.

Carbon Extras

To chemists and biologists, an "organic" substance is anything that contains carbon. Thus, living plants and animals, as well as their nonliving remains, are organic. To farmers and gardeners, "organic" means natural instead of man-made. To further complicate things, certain simple compounds which *do* contain carbon are called "inorganic" compounds.

Combustion, a fancy way to say burning, converts a fuel into heat and light. Combustion requires oxygen and releases other substances that were contained in the fuel. Once upon a time, combustion of carbon could occur in only three ways. Lightning could strike a tree and ignite it, thus releasing the carbon within the tree into the air. A volcanic eruption's flow of super-hot lava could burn the grasses and animals in its path, allowing their carbon to enter the atmosphere. Or a super-hot meteor falling to Earth could burn everything in and around the impact zone, and allow carbon to move into the atmosphere.

Carbon could also be released into the atmosphere by other natural processes. Long-buried items containing carbon were brought to the surface by earthquakes, or exposed to the atmosphere when wind and water wore away the soil or rocks above them.

Adding People to the Mix

Then people came along, discovering ways to change the natural sequence and timing of events. When humans began to manage fire, they created new and different ways to release carbon, ways that are faster, more frequent, and more widespread than natural events. Humans could start fires any time and any place—and they did, a lot, because fire is useful. Instead of lightning burning the occasional grassland or stand of trees, humans could burn whole fields and forests for clearing, planting, and building, instantly releasing carbon that might've stayed put for millennia, freeing it to move about and combine with other elements.

Things really took on a new speed and breadth when humans discovered how well coal—and then oil—burns, and they began mining and drilling beneath Earth's surface to find more of these *fossil fuels* to burn.

These are the principal fossil fuels humans use today:

- **Petroleum** is an oily, liquid solution of hydrocarbons. It can be changed into many fuels, such as gasoline, kerosene, diesel, and fuel oil. Also known as oil.

- **Natural gas** is a mixture of gaseous hydrocarbons, chiefly methane.

- **Coal** is formed from the ancient remains of plants. It can be used as fuel as is or can be changed into other forms, such as coke.

- **Coke** is coal with most of the gases removed. It produces a particularly intense heat that is useful in certain manufacturing processes.

def•i•ni•tion

Fossil fuels are organic substances usually found underground, where they developed during previous geologic periods, that are used as a source of energy.

Each of these fuels contains varying amounts of carbon. Peat is carbon on its way to becoming a fossil fuel, but it is relatively young and only partially decayed. Although not common in the United States, peat is a valuable fuel in other parts of the world.

When humans started removing fossil fuels from their long-hidden locations to use them for a variety of purposes, the opportunities for carbon to move around increased dramatically.

The Good Greenhouse

Now we'll take a look at some of the ways human use of fossil fuels changes Earth's natural systems. Frogs and ferns, ants and pansies, dolphins and daisies—the many living things here on Earth can survive only because our planet is surrounded by an atmosphere. This mixture of gases and water vapor nurtures and protects us.

The molecules of gases in Earth's atmosphere perform several functions. They trap some heat near the surface, which is good—otherwise, the planet might be too cold for plants and animals. They also act as a protective layer, strong enough to stop too much solar radiation from getting to the surface by reflecting it back into space. This is good, too.

The interplay between incoming energy from the sun and the gases in the earth's atmosphere makes life on our planet possible.

Carbon Impact

Earth's atmosphere extends far above the surface in five distinct layers, consisting mostly of nitrogen (78 percent) and oxygen (21 percent); the other 1 percent is a mixture of many different gases, within which the proportions of various gases do change. Scientists study how changes in these gases are linked to changes in global climate.

In the early 1800s, French scientist Jean-Baptiste-Joseph Fourier compared the blanket of gases in Earth's atmosphere to a greenhouse. In a real greenhouse, sheets of fine-meshed screen, plastic, or glass act as the roof and walls. They can be controlled to let in what's needed on any particular day, or to let out what's not needed, regulating the temperature, light, and even humidity within the structure. Controlling a whole planet's atmosphere is a lot more complicated. Let's take a look at what's in our atmosphere.

Earth's Greenhouse Gases and More

The most abundant gases in our atmosphere are nitrogen (N_2) and oxygen (O_2). You'll notice that each of these two gases occurs as two molecules stuck together. It's a very tight bond—so tight that, when heat comes along, these molecules don't wiggle, nor do they absorb the heat. So nitrogen and oxygen are *not* greenhouse gases.

The job of absorbing and then later radiating heat goes to a group of much less abundant—yet extremely important—gases. These are the three main greenhouse gases:

- Carbon dioxide (CO_2)
- Methane (CH_4)
- Nitrous oxide (N_2O)

The atoms in each of these gases are not bound to each other quite as tightly as those in nitrogen and oxygen. When heat comes along, these greenhouse gases can bounce and wiggle just enough to absorb some of the heat, and then release it again later. The cycle of heat absorption and release (also called radiation) is ongoing as these greenhouse gases continually pass energy among themselves.

 Carbon Extras

Each element has a distinct scientific symbol of either one or two letters of the alphabet. In many cases, the connection is easy to see—O stands for oxygen, H stands for hydrogen. In other cases, you have to know a bit of Latin—for example, gold's symbol, Au, comes from the Latin word *aurum*.

When you see a subscript (a number written on the right side and slightly below the element's symbol), the number indicates that more than one molecule of the element is present.

A scientist writes water as H_2O—two molecules of hydrogen grouped with one molecule of oxygen. In your bathroom cabinet, you might have a product containing hydrogen peroxide, H_2O_2, two molecules of hydrogen combined with two molecules of oxygen. Elements combine in particular ways due to their internal structure.

The three main greenhouse gases get help in regulating the temperature here on Earth from another very abundant gas—water vapor. When enough water vapor is present in a particular area, we can see it either as fog at ground level or as clouds in the sky. Water, our familiar H_2O, moves around as part of another planetary cycle as moisture travels from clouds to the surface as precipitation, then into rivers, lakes, oceans, or aquifers far underground.

The Greenhouse Effect

Solar radiation passes through the clear atmosphere.

Some solar radiation is reflected by the earth and the atmosphere.

Most radiation is absorbed by the earth's surface and warms it.

Some of the infrared radiation passes through the atmosphere, and some is absorbed and re-emitted in all directions by greenhouse gas molecules. The effect of this is to warm the earth's surface and the lower atmosphere.

Infrared radiation is emitted from the earth's surface.

How Earth's greenhouse gases trap some things and cause others to bounce away.

Another group of greenhouse gases usually gets lumped together under the heading fluorinated gases:

- Hydrofluorocarbons
- Perfluorocarbons
- Sulfur hexafluoride

You may see them abbreviated as CFCs or HCFCs. Generally, the fluorinated gases (hydrofluorocarbons, perfluorocarbons, and sulfur hexafluoride—also known as CFCs and HCFCs) consist of combinations that don't happen naturally.

Human technology started causing these combinations to form during the last century or so, and although these gases are not very abundant, they have plenty of impact on the way our atmosphere works. These gases affect ozone, another feature of Earth's atmosphere. (We'll take a look at ozone in Chapter 9 on carbon initiatives.)

Ebb and Flow

The greenhouse gases do not remain constant. Natural processes such as the growth and decay of plants and the breathing of animals mean that the amounts of greenhouse gases change every second. Scientists call this process *flux*.

def•i•ni•tion

Flux is the word scientists use to describe the way molecules move about, forming different combinations. You might think of it as the way the components flow from one place to another.

In the natural carbon cycle, things happen that tend to keep the flow of molecules in the carbon cycle truly a circulating pattern. Animals exhale carbon dioxide, then plants and the oceans tend to absorb that carbon dioxide from the atmosphere. The process repeats continuously everywhere.

When human activities cause more of a particular gas to appear than would be expected from the natural cycles already operating, the results get a special name. Scientists call these *anthropogenic emissions.*

def•i•ni•tion

Anthropogenic emissions occur when human actions release elements and compounds into the atmosphere.

Anthropogenic emissions can be grouped by the kinds of gases involved. Each kind of gas comes from many human activities. Carbon dioxide (CO_2) comes from ...

- Fossil fuel combustion (from fixed locations such as a manufacturing plant or electricity-generating station, and from mobile locations such as a car, boat, or airplane).
- Deforestation (when timber is harvested or forests are burned to clear the land for another use).
- Industrial processes (such as making steel or cement).

Methane (CH_4) comes from ...

- Mining, transporting, and burning fossil fuels.
- Burning other fuels, such as cornstalks or wood.

- Livestock.
- Rice cultivation.
- Landfills.

Nitrous oxide (N_2O) comes from ...

- Using fertilizers that contain nitrogen.
- Burning fossil fuels and wood.

Carbon Impact

The three main greenhouse gases, carbon dioxide, methane, and nitrous oxide, are chemically different and have different properties. But for convenience, emissions of these various gases are converted into one standard, that of carbon dioxide. That's why we call it your "carbon footprint," not your "carbon dioxide-methane-nitrous oxide" footprint.

A Little or a Lot?

At first glance, these lists would make you think just about anything humans do pumps huge amounts of gases into the atmosphere. But is it really that much? We've already learned that the greenhouse gases are the least abundant gases in Earth's atmosphere, each making up only a portion of that 1 percent that isn't oxygen or nitrogen.

Scientists are having a tough time collecting reliable data about the greenhouse gases, for many reasons. The greenhouse gases are dispersed throughout such a huge area that the only practical thing to do is take samples in small areas and then make estimates for the entire area.

Another problem is caused by seasonal variations. Remember that plants can absorb carbon dioxide into their leaves? Well, in the Northern Hemisphere, when trees drop their leaves in autumn, that interrupts the cycle of absorption until the following spring and changes the amount of carbon dioxide in the atmosphere for many months. And there could be other natural forces affecting the quantities of greenhouse gases in both the short and long terms. Scientists are not yet certain of all the effects that variations in the amount of energy radiating from the sun might have on greenhouse gases. Scientists also monitor the gases emitted during major volcanic eruptions to see what impact they have.

Carbon Extras

Although scientists now have many measuring tools for monitoring gases, figuring out what greenhouse gas concentrations were like 500; 5,000; or 5 million years ago presents its own set of problems. Scientists can use ice cores and study ancient layers of volcanic dust and other evidence for clues, and then make estimates.

Human activities that could be changing the natural ebb and flow of greenhouse gases are so varied and dispersed over the entire planet that they're hard to measure, too. But many nations have begun assembling lists of emissions of the top three gases—carbon dioxide, nitrous oxide, and methane—as well as other gases and tiny particles so they'll have a baseline to compare to in succeeding years.

Now that you understand how carbon in its many forms moves around in, on, and above Earth's surface, you're ready to start looking for the numbers that will help you measure your carbon footprint.

Doing Average Math

Has this happened to you? You read a statement such as "The average American eats more than 50 pounds of popcorn every year" and wonder, "Where did they get that number?"

Who is that ever-popular "average" American, and how can people say all those statistical things about him or her? One way to find an average is to take one "known" thing, such as the sales of unpopped popcorn, and then divide it by another "known" thing, such as the population of the United States. Some other limiting factors are usually involved: the numbers are from the same time period or pertain to a certain geographic area.

Popcorn is pretty easy to measure—you can pick up a single unpopped kernel in your hand, or help yourself to a nice serving of popped popcorn with your hand or a big scoop. You can see and feel and taste and smell it—it's right there in front of you.

Now suppose you wanted to find out what effect the average American has on greenhouse gases and the entire carbon cycle. A lot more variables are involved. First, you'd have to make a list of all the different activities that produce greenhouse gases. Then you'd need to figure out how much of each kind of greenhouse gas each activity produces, convert that to units of carbon dioxide, and then know the population of the United States—whew! The number and complexity of the calculations would take a long time to compute.

 Carbon Caution

> If you search the Internet for carbon-footprint calculators, you'll find plenty of choices. Some sites ask a bare minimum of questions and then give you a quick answer in just a minute or two. Other sites claim to be more accurate, with questions that take up to 15 minutes to answer.
>
> Even at these more detailed sites, your carbon footprint answer should be considered simply a very rough estimate. That's because the number is based on many averages and hidden assumptions that may not be accurate for your particular case.

Something else makes calculating your carbon footprint even trickier. It has two parts! The *primary* carbon footprint of an activity is what happens right now. For example, if you drive to the grocery, your primary carbon footprint consists of the greenhouse gases released while your car is running. Your *secondary* carbon footprint consists of the greenhouse gas impact of the things that are an *indirect* part of this activity. That could include how the parts of your car were made, then the energy impacts of the road's construction, how the food got to the grocery, and so on.

Since 1990, the United States Environmental Protection Agency (EPA) has been gathering statistics about greenhouse gas emissions from dozens of manufacturing sectors, agriculture, utilities, mining, and fossil fuel processors. The EPA's inventory, which now covers more than 15 years of data, converts the known greenhouse gas emissions in each category into one standard measure: units of carbon dioxide. This information is posted on the Internet and can be accessed by the public without a password. The United Nations has also been collecting greenhouse gas emissions data from many countries around the globe since 1990. This information is also available to the public on the Internet. The United States Census Bureau posts free data about human populations, business data, and other statistics at its website, too.

Anybody can take publicly available statistics from any of these sites, do a few calculations, and present the answer as a fact. Just remember that these much-manipulated numbers do not have the same clarity as a single measurement, such as the number of pages in your local newspaper this morning.

Carbon Caution

Evaluating information on websites can be challenging. As you read online, ask yourself two questions:

1. Who is responsible for the information on this website? The content on websites of universities and government agencies has generally been reviewed by experts before posting. There may be a bias, but it's usually toward conservative numbers and traceable documentation.

2. What other websites do they link to? A website that claims scientific objectivity may, after careful examination, reveal connections to a group with a specific political or social agenda; such ties could mean that data is far from neutral.

A special note: Trade association websites that present facts and figures from members who manufacture or sell similar products can provide excellent insights and fascinating details. But remember that each group is trying to project a certain image—examine claims carefully!

From Average to Unique

One common statistic that comes up over and over in discussions of carbon footprints claims that the average American is responsible for 20 tons of carbon dioxide entering Earth's atmosphere each year. But if you use public transportation during the work week and drive a car only on the weekends, and you don't enjoy flying and take the train only for business trips, and live in a region with very cool summers and don't need air conditioning, then you're not "average." Your situation is unique—and that's where this book can help you. Instead of trying to come up with nationwide averages, in the next chapters, we look at numbers that will have value in your particular case. A carbon footprint is definitely not one-size-fits-all!

The Least You Need to Know

- Carbon, in its many forms, is a natural part of Earth, living things, and the atmosphere. Human actions affect the carbon cycle.

- The greenhouse gases in the atmosphere make life possible here on Earth. Variations in the percentages of those gases have been linked to global changes in climate over time.

- Carbon dioxide, methane, and nitrous oxide are the three main invisible greenhouse gases. Water vapor, another component, is often visible as fog or clouds.

- Many human activities, including fossil fuel extraction and use, agriculture, and manufacturing, contribute quantities of greenhouse gases worldwide. Your carbon footprint is a way to estimate how your actions affect greenhouse gases.

Chapter 2

A Closer Look at Energy

In This Chapter

- An overview of the electrical grid
- Exploring the petroleum industry
- The natural gas network
- Energy use and public water supplies

Power lines, highways, railroads—we see them all around us, but we don't really look at them. Barges travel up and down our rivers; ships arrive at our ports. Beneath our feet, gas and oil pipelines and water lines provide many of the necessities of daily life.

Our American landscape is crisscrossed with overlapping networks that both supply and use energy around the clock. It's important to remember that each kind of fuel used in these various sectors has a different energy content and, when used, produces a different assortment of greenhouse gases. Understanding how this infrastructure works is important in seeing how your use of energy affects your carbon footprint.

In this chapter, we explore how these systems are organized, the fuels they use to supply the energy we want when and where we want it, and how the networks themselves consume energy.

Some Electricity Basics

Some quirky things are involved in electricity, but before we talk about them, let's review the basics. Traditionally, electricity is generated at very large power plants. From these giant power plants, the electricity flows through cables called transmission lines to smaller, local distribution networks with smaller lines, then on to the consumer. People refer to this network as the grid. It does not have a regular pattern such as the rows of boxes on a piece of graph paper; the lines of the electric grid go in many directions and branch out depending on geography and population density.

 Carbon Impact

During a recent year, more than 2.2 billion tons of carbon dioxide entered Earth's atmosphere as a result of generating electricity here in the United States.

To create electricity, most power plants use a spinning electrical generator called a turbine, a device with blades or rotors. As the turbine turns at high speed, it generates electricity. Some force must be applied to get the spinning started and keep it going. At the majority of American power plants, pressurized steam is used to turn the turbine.

But what makes the steam? In the United States, these are the most common fuels:

- Coal: 49.7 percent
- Nuclear: 19.3 percent
- Natural gas: 18.7 percent
- Petroleum: 3.0 percent

Electric Power Generation by Fuel Type (2006)

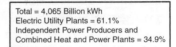

Total = 4,065 Billion kWh
Electric Utility Plants = 61.1%
Independent Power Producers and
Combined Heat and Power Plants = 34.9%

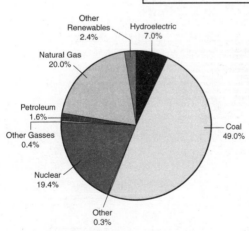

Electricity generation by fuel type.

Traditional, large-scale generating stations produce power measured in hundreds of megawatts. They operate continuously and use steady sources of fuel.

Three of these—coal, natural gas, and petroleum—are fossil fuels, which together account for 71.4 percent of electricity generation today. When burned, each of these fossil fuels releases a different combination of greenhouse gases.

Where a power plant is located and the kind of fuel it uses are closely related. A power plant that uses coal needs to be located near a railroad or a river where rail cars or barges filled with coal can make continuous deliveries. A natural gas power plant must be near gas pipelines. These kinds of conventional power plants are rather compact, occupying only a few acres to produce a lot of electricity.

 Carbon Caution

> Nuclear power plants do not use fossil fuels and do not emit greenhouse gases. However, nuclear power plants do produce radioactive wastes in the form of spent fuel, tools, protective clothing worn by the workers inside the plants, and many other items contaminated with radioactive materials. What to do to keep these waste products from harming people and the environment is a problem that remains to be solved.

Clean *and* Infinite

But there are other ways to generate electricity. The force of falling water can also provide the energy to turn a turbine. One of the most well-known and easily recognized such power plants operates at the base of Hoover Dam. Built in 1939, the power plant at Hoover Dam contains 17 main turbines. Using the force of falling water skips the step of having to create steam and reduces emissions of greenhouse gases. In the United States today, only a small portion—just 6.5 percent—of our nation's electricity is generated using the force of falling water at hydroelectric plants. An even smaller portion—just under one-half of 1 percent—of our nation's energy comes from wind power.

Wind power plants (often called wind farms because many wind machines are grouped together) also skip the steam step and channel the energy of the moving air directly to turbines to generate electricity; no other fuel is required. The Horse Hollow Wind Energy Center in Texas contains 421 wind turbines installed on 47,000 acres in Taylor and Nolan Counties in Texas.

Using the sun to generate electricity also requires no fossil fuel. The energy of the sun is captured in a solar-thermal device that concentrates the heat and then transfers it to a liquid. Through other steps, steam is eventually produced to turn turbines, just like in a conventional power plant. The Solar Electric Generating System in San Bernadino County, California, is an array of solar panels at several distinct sites scattered over 40 miles.

The fossil fuels used to generate electricity are nonrenewable. There is a fixed supply, and once it's used up, it's gone. But these other kinds of fuels for generating electricity are renewable: the energy that comes from water, wind, or the sun, or from something that can be replenished on a regular basis, such as a crop or waste products from some other activity. In general, renewable energy sources tend to release smaller quantities of greenhouse gases than nonrenewable sources. However, they may not produce equivalent amounts of usable energy from the same volume of fuel, or they may require a much larger physical space than a conventional power plant.

Do you know whether your electricity comes from using fossil fuels or from renewable sources? Most likely, your electricity comes from a combination of sources. To find out how the electricity in your area is generated, check the website of your local electric utility. If you want more detailed information for your region, you can try searching the Internet by using the name of your state along with the words "electricity" and "generation."

Measuring Power

The basic unit of measurement in electricity is the *watt*. Using electricity is measured in terms of watts and hours. A watt is a very tiny amount of electricity, so most of the time you will see references to *kilowatt-hours*, abbreviated as kWh. A kilowatt is the same as 1,000 watts.

One letter in an abbreviation makes a big difference. From smallest to largest, here are some ways you might find quantities of electric energy described:

- Watt-hour = wh = the basic unit of measurement
- Kilowatt-hour = kWh = 1,000 watt-hours
- Megawatt-hour = mWh = 1 million watt-hours
- Gigawatt-hour = gWh = 1 billion watt-hours
- Terawatt-hour = tWh = 1 trillion watt-hours

def•i•ni•tion

A **watt (w)** is a measure of electric power. One watt equals $\frac{1}{746}$ horsepower. Many appliances and devices are labeled with the number of watts they need to operate. Examples are a 100-watt light bulb, a 1,500-watt hair dryer, and a 120-watt stereo system.

One kilowatt-hour is the amount of electricity needed to power a single 100-watt light bulb for 10 hours.

Managing Supply and Demand

Due to the physical properties of electricity, the supply (what is generated) must match the demand (what users need) every second around the clock. Electricity flows through the grid continuously and must be used as it is generated. Even the largest batteries can store only relatively tiny amounts of electricity for use at another time.

The demand for electricity in any geographic area varies according to the time of day or night, the time of year, and many other factors. Power plant operators use historic data, weather predictions, population data, and many other statistics to predict how much electricity will be needed at any particular time, and then they generate precisely that amount.

Peak demand is when the most electricity is needed. Power plant operators call this the peak load. During off-peak demand or load times, less electricity is needed. A utility may have enough generating capacity at its usual power plants to meet off-peak demand needs with its usual methods of generating electricity. But in times of peak demand, it may need to increase capacity by using additional power plants or by obtaining the extra electricity from another source in another area.

Alternative Fuels and Methods

At conventional large-scale power plants, a steady source of fuel can be relied on to meet the usual demand for electricity. But today a gradually increasing number of smaller-scale power plants use different kinds of fuel:

- *Biomass*
- Methane from landfills
- Methane from livestock

def•i•ni•tion

> **Biomass** refers to plant material used as a fuel. Sometimes it's as simple as wood logs for a fire. Or biomass can be leftovers from another activity. Bits of bark and sawdust at a paper mill are biomass; agricultural waste products, such as the stalks and leaves of corn plants that remain after the ears of corn have been harvested, are another form of biomass.

Other kinds of smaller-scale power plants may use energy from the sun or energy from wind. They are just much smaller versions of the ones mentioned earlier in this chapter. However, even grouped together, these uncommon fuels generate less than 3 percent of the electricity in the United States today. That percentage is expected to increase; we look at that in Chapter 9 when we look at carbon initiatives. When utility companies build and own these renewable energy source–generating facilities, they're connected to the grid in the normal way.

But an important innovation concerns the smaller versions of power plants that use these renewable forms of energy to generate electricity. When individuals or businesses produce electricity, it's called *distributed generation*, meaning that it takes place somewhere other than at a centrally located, conventional generating plant. These mini power plants are distributed throughout the entire grid and very often are located close to where the power can be used. The electricity may be used on the spot with no connection to the grid, or the mini power plant may be connected to the grid so that any surplus can be added to the traditional distribution system.

At industrial and manufacturing plants, independent electricity supplies are often created using cogeneration and other recycling or dual-use methods. *Cogeneration* can mean two different things. In one situation, cogeneration means using a waste item, such as tree bark at a paper mill, as the fuel to generate electricity. In other situations, cogeneration means to use a fuel to produce electricity and something else useful, such as steam or heat. As distributed generation and cogeneration become more common, they may help reduce greenhouse gases significantly.

Petroleum Industry Basics

After electricity generation, the biggest contributor to greenhouse gas emissions in the United States involves our use of petroleum. Most of that petroleum goes to providing fuel for transportation.

Crude oil, another name for petroleum, must be refined or separated by a system of chemical and mechanical processes into various useful products. Petroleum is the basis for plastics, synthetic fibers, tires, crayons, medicines, and dozens of other products, but the largest percentage of petroleum products are fuels:

- Gasoline
- Diesel
- Jet fuel
- Aviation gas

- Kerosene
- Heating oil
- Liquid petroleum gas (LPG or LP gas)

Carbon Dioxide from Various Fuels

One Gallon of This Fuel	Pounds of CO_2 Produced
LP gas (liquefied petroleum)	12.805
Aviation gas	18.355
Motor gasoline	19.564
Jet fuel	21.095
Kerosene	21.537
Distillate fuels (home heating oil, diesel)	22.384

Crude oil arrives at a refinery by either a pipeline or a ship. After the various substances are separated during the refining process, they leave the refinery according to the most appropriate shipping method for each product. The refining process itself uses energy, and each of these transportation networks uses still more energy. These petroleum products could go into different pipelines, ships, tanker trucks, or barges, or some combination of methods, to get to their eventual destinations.

Carbon Impact

The tanker truck that you see delivering gas to your local station is just one part of a very long journey that uses energy for transportation before the fuel ever gets to your vehicle.

Before you look at the table of the carbon dioxide emissions from various fuels, you have to understand a little bit of physics. Remember in Chapter 1 when we described molecules? Each kind of molecule has an atomic weight. Molecules are so small they don't feel heavy to us. You wouldn't feel the weight of one single molecule in the palm of your hand. But a gallon of gas sure is heavy, something you know if you've ever had to lug one back to your car stranded on the side of the road!

A gallon of gas (or any other kind of fuel) contains lots of carbon molecules, each one with an atomic weight of 12. The amount of air (a mixture of many gases) needed for combustion contains lots of oxygen molecules, each one with an atomic weight of 16. Carbon dioxide contains one carbon molecule and two oxygen molecules, for a combined atomic weight of 44. So even though it sounds strange, burning a gallon of gas that weighs about 7 pounds produces almost 20 pounds of carbon dioxide! That's because of the added weight of the oxygen from the air that was part of the combustion process.

We'll explore how using fuels for vehicles affects your carbon footprint in Chapter 5.

Natural Gas Basics

In everyday language, natural gas is methane that we use as a fuel for furnaces, power plants, and other purposes. *Natural gas*, like coal and petroleum, is a fossil fuel found underground. This gas is naturally odorless and colorless—the smell you notice when there's a leak comes from an artificial product added to make it possible to, well, smell a leak.

The raw natural gas that comes from underground or beneath the ocean is usually a combination of different things that can be separated into three main products:

- Methane
- Butane
- Propane, which, when liquefied and stored under pressure, is known as LP gas or LPG

def•i•ni•tion

Natural gas has two meanings. In the first sense, it is a catch-all term for the raw gas that occurs beneath Earth's surface. In its other sense, natural gas means the purified form of that gas consisting of just methane.

Did you notice something odd? As you learned in Chapter 1, methane is one of the principal greenhouse gases. So how does using natural gas, which is methane, affect the emissions of greenhouse gases?

First, throughout the processing and distribution of natural gas, great care must be taken so that very little of this methane escapes directly into the atmosphere. Second, although burning natural gas does produce carbon dioxide, it produces very few ash particles. Burning natural gas also emits less nitrogen, sulfur, and carbon than oil and coal emit.

Most of the natural gas used in the United States comes from wells either on the continent or off-shore; some is imported from Canada. Unlike electricity, natural gas is relatively easy to store until it is needed. The distribution network for natural gas consists of several parts:

- Underground storage near the producing fields
- Large pipelines for long distances
- Smaller pipelines called "mains"
- Even smaller pipelines called "services" that go to the consumer

Most natural gas is used for winter heating purposes, so during the summer months as natural gas is produced, it goes into temporary storage. Then during the winter months, it can be withdrawn from storage and fed into the pipelines as needed. During the hot summer months, natural gas can be used to generate extra electricity to meet peak demand in southern states.

The physical properties of butane and propane mean that each of these forms of natural gas must be handled and distributed differently. We'll explore the carbon impacts of these other products in later chapters.

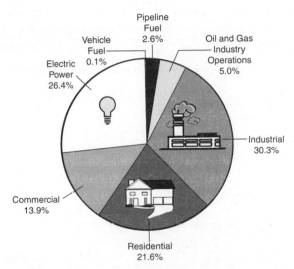

Natural Gas Use

Pipeline Fuel 2.6%

Vehicle Fuel 0.1%

Oil and Gas Industry Operations 5.0%

Electric Power 26.4%

Industrial 30.3%

Commercial 13.9%

Residential 21.6%

Natural gas usage.

Water, Water Everywhere, and Sewage, Too

Every school kid knows that water and electricity is a deadly combination. But electricity is used to power pumps and other equipment throughout the nation's water supply.

Purifying water for human consumption and then treating the wastewater afterward consumes energy at a surprising rate. A gallon of water is heavy—about 8.3 pounds. It takes a lot of power to move water even a short distance. Most of us think of water for household uses, such as bathing and cooking, but it's also a key part of many manufacturing processes. How and when you use water affects your carbon footprint in some obvious and more subtle ways.

In our nation's water supply, energy is needed to …

- Move water from its source (river, lake, catch basin, reservoir, or underground aquifer).
- Chemically and physically treat the water to make it safe for humans.
- Move the conditioned water through pipes to users.
- Heat the water for certain uses.
- Move the used water to the wastewater treatment plant.
- Operate the wastewater treatment plant.
- Disburse the treated wastewater for reuse or return to the natural system.

Water towers, either freestanding ones or large tanks installed on rooftops, play an important role in maintaining the pressure and supply of water to communities. Communities and buildings choose the size of the water tower or tank so that it will hold about one day's supply of water. Using a pump strong enough to meet the average demand for water throughout the day, and then adding some of the water stored above ground in the tower or tank by simple gravity during times of high demand, allows the utility to maintain normal pressure and flow. And it means that the utility does not have to invest in a much more expensive and more energy-consuming pump that would be needed to supply the larger volume of water during times of peak demand.

Wastewater treatment facilities rely more on gravity than freshwater supply systems do, but they still use energy during several steps of the process.

How America Uses Water

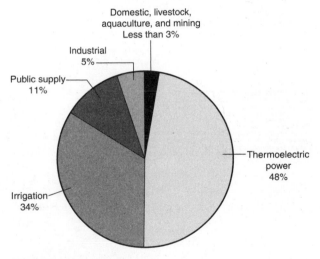

How America uses water.

When you use water wisely, you reduce the need for additional pumping and other energy required throughout the water system. That reduces your carbon footprint. We'll look at many ways to reduce your carbon footprint through careful use of water in many sections of this book.

The Least You Need to Know

- All the various methods used today to generate electricity in the United States contribute more than 2 billion tons of greenhouse gases to Earth's atmosphere each year.

- Refining, distributing, and using petroleum products is the second largest source of greenhouse gases emitted in the United States.

- Natural gas is pure methane, one of the greenhouse gases. Burning natural gas produces carbon dioxide, another greenhouse gas.

- Electric power plays a vital role in all parts of our nation's fresh water supply and wastewater treatment facilities.

Chapter 3

Carbon at Home

In This Chapter

- Heating and air conditioning—the home energy guzzlers
- Windows, doors, and insulation
- Using water wisely
- The Energy Star program

Americans move around a lot. So far, I've lived in four apartments, two townhouses, and five different single-family houses. My opportunities to control energy usage have been very different in each place I've called home.

In this chapter, we take a look at the areas within any home (whether it's a rented apartment, a mobile home, a condo, or a free-standing house) that affect your energy use. Remember that "average American" in Chapter 1? One study showed that, on average, about 50 percent of energy use within our homes is for heating and cooling, about 30 percent for appliances and lights together, and about 20 percent for heating water.

Your energy use is probably different, but these everyday areas provide a good starting point to look at how you can reduce the size of your carbon footprint. Depending on your current situation, you might be able to make some big changes or just a few little changes now to reduce your carbon footprint. If you're planning to move soon, you'll have a head start on knowing what to look for as you evaluate your energy options at your next place.

Heating, Cooling, and Ventilating Systems

The key to reducing your carbon footprint at home involves two steps—choosing the most energy-efficient home systems and appliances possible, and then using them responsibly to prevent waste. The energy you use to make indoor spaces comfortable accounts for a huge portion of your carbon footprint.

The general term *HVAC* refers to all the parts of the system used for heating, ventilating, and air conditioning a building. It includes the equipment that produces heat or cooling, the way that changed air is disbursed through the building, how the air is modified (such as adding or removing humidity or removing contaminants), and how fresh air enters and leaves the system.

def•i•ni•tion

> HVAC (heating, ventilating, air condition-
> ing) stands for the system or systems that
> heat, cool, and provide fresh outdoor air
> to the interior of a building. These systems
> also often change the moisture content of
> the air, as well as purify it by removing
> contaminants or irritants.

The kind of heating system your home has depends
on a complex set of evaluations. HVAC professionals in your area have already worked out the best
options for your climate zone by examining the
cost and availability of fuels in your region, the
installation and maintenance costs, the way your
home is constructed, and whether the same system
will be needed for cooling during the summer.
Several options might work, and what system you
have often depends on how old the property is.

 Carbon Extras

> In many cases, the most energy-efficient
> choice may involve using zones and
> auxiliary items. You might save energy by
> adding a small item such as a fan or room-
> size space heater to keep one zone of
> your home comfortable when people are
> in that area, and then turning it off when
> the space is empty.

Important ways you can affect how much energy your existing HVAC system uses include these ...

- Using your thermostat properly.
- Maintaining proper humidity levels indoors.
- Controlling how outdoor air contacts indoor air.
- Improving insulation.

If you're planning a major remodeling project or building a home from scratch, be sure to ask your contractor to consider innovative energy-saving designs. He or she may be unfamiliar with some of the newer concepts and be reluctant to discuss them. Shop around. You could get lucky and find a contractor who will turn your project into a show-case for the latest energy ideas for your community.

Let's take a look at these options.

Understanding Thermostats

Setting your thermostat to a lower temperature during winter means you will use less energy. Setting your thermostat to a higher temperature during summer means you will use less energy. In an older home or apartment, you may have a simple thermostat that you control manually. You can set it to a different temperature before you go to bed, then you must reset it at breakfast, and then you must remember to reset it again before leaving for work.

Many newer homes, condos, and apartments feature a programmable thermostat that allows you
to preset higher and lower temperatures at various times of the day and night, and on weekends.
These can save a lot of energy because the changes
take place automatically—they never forget! They
do include override options so that you can always
make adjustments for special occasions such as
holidays and vacations.

A heat pump is an electric device that can provide
heating in winter and cooling in summer. There
are many different kinds of heat pumps. Electric
heat pumps work most efficiently in winter when
the outdoor temperature is between 45°F and
50°F. When the outside air is colder than that, an
auxiliary system that uses more fuel is needed to
maintain comfortable indoor temperatures. But
beware! During the winter, if you set an older
thermostat to too low of a temperature, moving
the thermostat to a warmer temperature may cause
your backup heating system to work harder and use
more energy in the long run. How long you leave
your thermostat at the lower temperature may be
a factor. Check with your local HVAC professional
and consider installing a special "setback" thermostat designed for use with heat pumps.

Humidity Makes a Difference

Changing the humidity of air affects the way
we humans perceive the temperature. Removing
humidity from your indoor air makes it seem
cooler, even though the real temperature is higher.

Most central air-conditioning systems have a dehumidifying function built in; you cannot control how much moisture is removed from the outside air that flows into the system.

Adding humidity to your indoor air will make it seem warmer, even though the real temperature is lower. In the winter, you can control how much humidity is added to your indoor air two ways. If your HVAC system has a humidifier attached to the furnace, you'll have a control feature or dial (usually mounted near the heating system's thermostat) that allows you to choose the percentage of moisture. Most also include recommendations about the proper percentage to maintain according to what the outside air temperature is.

You can also add humidity to your indoor air in winter the old-fashioned, low-tech way, with shallow containers of water sitting on or near your heat vents or radiators. But *never* place water containers directly on an electric device or appliance.

 Carbon Caution

The greater the difference between the outside and inside air temperatures, the more potential there is for condensation to form on window glass and other surfaces. Every home is different, so you may need to experiment a bit until you get just the right combination for comfort.

Keeping the Outside Out and the Inside In

It's one thing to talk about the amount of fresh air that needs to circulate within a building as part of the HVAC system. Such ventilation is crucial for health and safety. But what about air that moves in or out due to gaps in the building's *envelope?* How important is that to your carbon footprint?

def•i•ni•tion

Engineers use the word **envelope** to describe the whole structure of a building; it's everything that separates the interior of the building from the outside elements.

Very! When air you've used energy to change escapes from your home, it wastes energy. When outside air enters your home, it makes your HVAC system work harder, and that wastes energy, too. Both situations cause your carbon footprint to be larger than it should be.

Four areas make up a building's envelope:

- Doors, windows, and skylights
- Wall systems and insulation
- Foundation
- Roofing

The windows, doors, and skylights in your home affect the interior temperature in three ways. Sunlight can enter through them, adding heat.

The materials of these items can gain or lose heat just by their contact with the outside. And air can leak in either direction due to gaps in the way these items are joined to the walls, roof, or foundation. Caulking (around the fixed elements of windows and skylights) and proper weather stripping to seal up gaps around moveable objects such as doors and window sections can save energy right away.

A window is only as energy efficient as the material it's made of and how well it's installed. The materials used to construct windows, their placement in your home relative to the position of the sun, and any interior and exterior coverings all make a difference in your energy use. If you want to improve the energy efficiency of your existing windows without replacing them, you can modify them from the inside, outside, or both. Here are some options that can reduce your carbon footprint:

- Install storm windows or plastic sheeting outside.

- Install exterior shutters that are operable instead of only decorative.

- Install exterior awnings, either fixed or retractable.

- Install plastic sheeting or films on the inside.

- Install window shades, blinds, or curtains on the inside.

- Make sure weather stripping is properly fitted. If you can pull a dollar bill through, the seal may not be tight enough.

- Try a no-cost, temporary solution to drafts and air leaks: a rolled-up towel laid across windowsills and thresholds.

Carbon Caution

Did you know that modern windows include insulation in two places? The frames may be made of different layers, perhaps metal over wood, to increase the window's insulating properties. The panes you look through may consist of a sandwich of special gases sealed between two sheets of glass; this layering effect also adds insulating properties.

You can determine the energy efficiency of a window by several rating systems. The two most important ratings are a window's *U-factor* and its *SHGC* number. If you are considering replacing old windows with new ones, you'll want to ask your builder or supplier for more information.

def•i•ni•tion

A window's **U-factor** measures how well it stops heat flow. A lower number means it does a good job. **SHGC** is short for "solar heat gain coefficient," which measures a window's ability to permit solar heat gain or loss. Whether you need a high SHGC number or low SHGC number depends on your climate, the location's exposure to the sun, and other factors.

Insulation Information

You know that hot air rises—that's because air is a gas. But heat moves differently through solids. Heat radiates in all directions in solids, and heat moves toward cold areas. This means two things for your living space. In the winter, the warmth in your home constantly moves toward the cool spots: floors, walls, ceilings, and windows. In the summer, hot outdoor air warms the exterior of your home; that heat then moves toward the cooler indoor materials. Anything you can do to slow down or stop the movement of heat in either direction improves the efficiency of your HVAC equipment.

The better your insulation, the lower your carbon footprint will be in every season. If you're a renter or condo owner, you can't do much to change the materials used to construct walls or the kind of insulation used in your building. But if you're a homeowner, you have many choices.

To lower the amount of energy you use to heat or cool your home, you can increase your insulation in these areas:

- Exterior walls—even walls that border your garage
- Attics and all spaces between your roof and living areas
- Crawl spaces and unfinished basements

Carbon Caution

> Your home's exterior and the insulation must work together to control moisture. Wet insulation is not as effective as dry insulation, so insulation must be installed carefully.

Did you know that many air leaks can be traced back to the spaces around electrical outlets and where pipes come into your house? Special insulation kits available for sale at home improvement stores and on the Internet can help you stop these tiny yet energy-wasting gaps in your home's insulation.

The *R-value* of a particular kind of insulation gives you information about how good of a job it does slowing down the flow of heat. The R-value of a substance depends on these factors:

- The physical properties of the material
- The thickness of the material
- The density of the material

def•i•ni•tion

> The **R-value** of a material measures its ability to resist the flow of heat. Insulation materials with high R-values are more useful as an energy-saving choice. When figuring the R-value of layers of different materials, you get to add them all together for a grand total.

Water Heaters and Lighting

In Chapter 2, you discovered that every time you use water, you affect the size of your carbon footprint. That's because it took energy to get that water to you, and it will take some energy to move it along to the wastewater treatment system. When you use heated water, you add even more to your carbon footprint because of the energy it takes to heat the water. Here are four ways to lower your energy use for heating water.

1. Lower the thermostat on your water heater.

2. Put a special safety-rated insulating "blanket" around your water heater to improve its ability to hold the heat.

3. Use special insulating wrap on your hot water pipes.

4. Replace your old water heater with a newer, more energy-efficient model (this is especially true if your old water heater has a pilot light).

Keep two things about water in mind to reduce your carbon footprint: use less water, and use less heated water. Reduce the total amount of water you use by ...

● Fixing leaking plumbing fixtures.

● Using less water in toilets.

● Turning off the water while you brush your teeth.

- Waiting to run the dishwasher and washing machine until you have a full load.

Carbon Impact

Can't replace a 5- or 3-gallon per flush toilet fixture? You can still save a lot of water with this simple, no-cost trick. Add a few inches of pebbles or sand to a used plastic drink bottle, fill the rest of the way with water, put the cap back on, then sit it inside a corner of your tank away from the mechanism. Now each time you flush your tank will refill with less water. You may be able to add a second bottle in another corner, and still have enough water to flush.

Reduce the amount of hot water you use by …

- Taking shorter showers.
- Washing laundry in warm or cold water as much as possible.

For more ideas about using water wisely, visit: www.epa.gov/watersense/index.htm.

The Energy Star Program and Other Rating Systems

When it's time to replace an old appliance or update some other part of your home, you may be surprised by how many labels these items have.

Each category of item, from washing machines to windows, to insulation, to furnaces, has standards that make it possible to compare items within the category to choose one that will suit both your energy budget and your dollar budget.

The *Energy Star* system of rating and labeling things that use energy has been helping consumers make better choices and reduce their carbon footprint since its introduction in 1992. The program, a joint effort of the U.S. Environmental Protection Agency and the U.S. Department of Energy, is voluntary. Manufacturers and builders submit their products for evaluation against rigorous standards. Items that earn the right to display the Energy Star label meet strict energy efficiency guidelines.

def•i•ni•tion

Energy Star is a voluntary government program that certifies that an item meets high standards for energy efficiency.

Thousands of products, from appliances to light bulbs and computers to entire houses, are now certified as energy efficient. For details, visit www.energystar.gov.

U.S. electric cooperatives (the utilities that provide electricity in areas not served by public or investor-owned utilities) have recently introduced the Touchstone Energy–certified home program, another rating system for energy efficiency. For details, visit www.touchstoneenergysavers.com.

 Carbon Impact

When you decide to buy and install any kind of product to save energy, save your receipts for whatever home energy improvements you make, even if you rent. Many states and the federal government offer incentives such as rebates and tax credits for improving energy efficiency. Consult your tax advisor for details.

On many products, you will also see a label that says "EnergyGuide." This label gives you important information that can help you reduce the size of your carbon footprint. The label includes …

- An estimate of the energy needed to operate this item during a full year.
- A comparison of the energy use of similar models.
- Where this particular item's energy use occurs on a scale from least energy use to highest energy use.

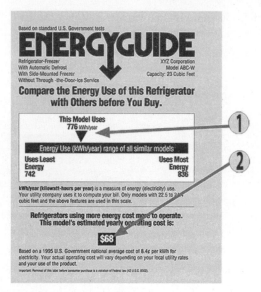

An EnergyGuide label.

In the rest of this section, we take a look at some of the terms to watch for when comparing products.

The opportunities to change your HVAC equipment are relatively rare, for two reasons. Usually whatever system is in place is an integral part of the building's structure. Also most HVAC equipment has a relatively long life; many have 10-year and longer warranties, with useful life spans of up to 20 years. Spare parts are often available for another 10 years beyond that.

You can evaluate the energy efficiency of your current furnace or compare new furnaces by checking the equipment's AFUE (Annual Fuel Utilization

Efficiency). A rating of 90 means that 90 percent of the fuel provides warmth, while the other 10 percent escapes as exhaust or is otherwise wasted.

Air-conditioning systems are labeled with a SEER rating. Air-conditioning systems (not window units) manufactured before 2006 had to meet a SEER rating of 10. Effective January 2006, the SEER rating for newly manufactured systems was increased to 13. A 13 SEER system is 30 percent more energy efficient than a 10 SEER system.

SEER stands for "seasonal energy efficiency ratio." Regulated by U.S. law and the U.S. Department of Energy, a central air-conditioning unit's total cooling output in British thermal units (Btu) during its normal annual usage period is divided by its total energy input in watt-hours during the same time period. A British thermal unit (Btu) is the amount of heat needed to raise the temperature of 1 pound of water by 1°F. The output of furnaces and air conditioners is measured in Btus.

In many parts of the United States, burning logs or other natural fuels in a fireplace or stove is a common wintertime heating choice for a whole house or as a supplement to other HVAC systems. Through an unusual bit of logic, this is considered a part of the natural carbon cycle. This form of energy use is seldom mentioned as having a significant impact on greenhouse gas emissions. In areas where fallen timber is a plentiful local resource, burning logs can actually produce fewer greenhouse gas emissions because it avoids using the transportation and distribution networks of other fuels.

Carbon Impact

If you want to make sure you're getting the most heat safely from your wood-burning fireplace or stove, visit the U.S. Environmental Protection Agency's website for information: www.epa.gov/woodstoves/index.html

Throughout the rest of this book, you'll find more information and tips to conserve energy.

The Least You Need to Know

- Changing indoor air (whether it's for winter heating, summer cooling, or both) typically uses the most energy in a U.S. home.

- Improving the energy efficiency of heating and air-conditioning systems, combined with stopping waste through appropriate insulation, can dramatically decrease your carbon footprint.

- Using less hot water and less water overall has a large impact on decreasing your carbon footprint.

- Energy Star labels and other rating systems give you the information you need to make energy-efficient choices.

Chapter 4

Carbon Away from Home

In This Chapter

- Improving energy efficiency in schools
- Green buildings and green roofs
- Carbon-friendly business travel

In a previous chapter, we looked at energy use at home—but have you thought about how much energy you use when you're not at home?

We go to school and work, enjoy sports and other leisure activities, attend religious services, and volunteer in community projects. Our carbon footprints go along with us everywhere we go and in everything we do.

In this chapter, we take a look at two areas that account for the largest amounts of time we spend away from home, our schools and workplaces. Whether you're a student or a teacher, the big boss or an employee, this chapter will give you a lot of practical ideas to help you and your community use energy more wisely.

A Big Energy Lesson

More than 76 million students go to public school in the United States, from preschool kiddies to young adults in college. If you aren't in school yourself right now, chances are, you or someone you know has children or grandchildren attending classes. One of your co-workers might be going to night classes. You might attend sports events or community and civic activities at your local school when the regular school day is over.

Throughout the United States, local governments own more than 135,000 buildings used for kindergarten through high school education. Add in the buildings owned by private schools, colleges, and universities, and you can see that they account for a very large portion of all the kinds of energy used in our nation. A recent survey estimated these percentages for the energy use in an average public school building:

- Heating and air conditioning, 52 percent
- Water heating, 20 percent
- Lighting, 19 percent
- Refrigeration, office equipment, and other, 9 percent

Just as there are no average people, there aren't any average schools—the school nearest you probably uses energy in quite different percentages. But experts believe that about one fourth of the $6 billion (with a *b*) that taxpayers spend on energy use

in America's public schools is wasted. Improving
energy efficiency in all schools would save a lot of
money and could reduce greenhouse gas emissions
by a significant amount.

Classroom Activities and Contests

Implementing energy-saving ideas in schools is
moving forward with a variety of approaches. Many
programs get the students involved as they put the
factual information they learn in science classes
into real-life applications. The success of these
programs depends on teachers and administrators
sharing the students' enthusiasm for energy-saving
changes, whether they're simple, low-cost changes
or projects requiring larger investments.

 Carbon Impact

Teachers looking for lesson plans and
other classroom energy ideas can get
started at www.eere.energy.gov/
education/lessonplans/, a website maintained by the U.S. Department of Energy.

From modest beginnings in 1980, the National
Energy Education Development (NEED) Project
has grown into a far-reaching network of teachers,
students, business leaders, and assorted government
agencies sharing energy information in practical
ways in all states. Individual states have been busy
developing lesson plans and other activities for all

ages which are typically offered under the heading "K–12 Energy Education Programs," commonly called KEEP.

The National Energy Education Development (NEED) Project offers students and teachers a wide selection of classroom materials adapted to each grade level, along with special projects that explain energy issues while encouraging community involvement. For more information, visit its website at www.need.org/.

Many national and state organizations sponsor annual student contests for the best ideas in energy conservation. Many individual states also offer their own version of materials that raise energy awareness for students of all ages, such as Wisconsin's K–12 Energy Education Program (KEEP). You can find items of particular interest to your state by searching the Internet using the name of your state followed by the phrase "energy education."

You, your company, or your civic organization can raise awareness of the need for energy efficiency in your community's schools by sponsoring a local contest. Here are some energy contest ideas:

- Best poster encouraging energy conservation
- Best essay on "How I Spent Less Energy This Summer"
- Best energy-saving project under $25

Better School Buildings

Constructing a new school offers a community the opportunity to build in energy savings from the ground up. The latest high-tech, energy-efficient strategies include these:

- Placing the building on its site to take best advantage of sunlight and local weather patterns
- Including both passive and active solar energy systems
- Positioning windows and skylights properly
- Adding appropriate insulation
- Installing Energy Star high-efficiency HVAC equipment
- Installing Energy Star office equipment and computers
- Using water-saving plumbing fixtures

 Carbon Extras

October is Energy Awareness Month—a great time to begin an energy-saving project at your school or business.

If a member of your family is a student, you can start reducing carbon footprints now by …

- Encouraging teachers to print handouts and tests on both sides of the paper.

- Installing paper-recycling bins in classrooms and offices.
- Setting up a clothing-recycling program for school uniforms.

Transportation Savings

School buses offer another area with great potential for reducing carbon footprints. Two important innovations help reduce greenhouse gas emissions by school bus fleets:

- Alternative fuels
- Advanced vehicle technologies

As we'll see in a later chapter, one of the problems with using alternative fuels is that they may not be readily available to the public in all locations. But since school buses travel regular routes and often return to central compounds, refueling with something other than common diesel can be easily arranged.

One new bit of technology is aimed at reducing the amount of time school buses sit idling. Idling an engine while waiting to pick up or discharge students contributes additional greenhouse gases to the atmosphere and wastes fuel. New devices for school buses can keep important gauges and systems working even when the engine is turned off. Ask your local school to make "no idle" zones for buses and other vehicles, to reduce emissions and conserve fuel.

Here are some ways to make getting to and from classes and other school events more energy efficient:

- Ride the school bus or use other public mass transit.
- Walk or bicycle part of the way when and where it's safe to do so.
- Carpool with neighbors.
- Arrive a little earlier in the morning or later in afternoon, to cut down on energy-wasting traffic jams around the school building.

Energy at Work

There are more than 4 million commercial buildings in the United States, from restaurants to car washes, dry cleaners to department stores, strip malls to high-rise office towers, to lawnmower repair shops. Add in industrial operations and manufacturing plants, and it's easy to see how business activities can account for almost half the energy used in the United States. Tremendous opportunities abound for improving energy efficiency and reducing carbon footprints in all these workplace situations.

These eight American industries use the most energy:

- Aluminum
- Chemicals
- Forest products

- Glass
- Metal casting
- Mining
- Petroleum refining
- Steel

Several major efforts have already made significant progress in helping businesses use energy more efficiently.

Managing Energy Within the Workplace

Workplaces where things are manufactured can include very complex interlocking functions—and tremendous potential for energy savings through such ideas as using waste heat from one process to heat water or provide cooling and space heating in another area. Experts in best practices (engineers who figure out the most cost-effective and least wasteful approaches to activities) are constantly coming up with innovations that save energy and are environmentally friendly.

At the U.S. Department of Energy (DOE), a special division called Energy Efficiency and Renewable Energy (EERE) offers a Building Technologies Program with practical ideas for existing structures and new construction in a variety of workplace situations.

In another DOE program, Save Energy Now, this government agency can conduct an energy assessment at a manufacturing plant or other industrial facility. These assessments review everything from heat used in manufacturing processes to fans, motors, conveyor systems, compressed air systems, and pumps, and then provide recommendations on ways to better manage energy use.

To find out if this program could help your business reduce its carbon footprint, visit www.eere. energy.gov/industry/saveenergynow/.

Office buildings present different opportunities to use energy more carefully. The energy use in a typical office building is allocated in this way:

- 34 percent for HVAC
- 30 percent for lighting
- 16 percent for office equipment
- 9 percent for water heating
- 11 percent for all other miscellaneous uses

Here are some quick strategies for reducing energy use in an existing office building:

- Turn off lights when an area is not in use.
- Install motion detectors to make turning off lights in unoccupied areas automatic.
- Install water-saving devices in all plumbing fixtures.
- Use programmable thermostats to save energy at night and over weekends.
- Set up and follow a regular maintenance schedule for all HVAC equipment.

And how about a list of things you can recycle at work?

- Office paper
- Ink cartridges from copiers and printers
- Cardboard boxes
- Packaging and shipping materials

Another key element in the movement toward increased energy efficiency is the concept of *green buildings.* These structures (which include offices and other workplaces, homes, and schools) and their construction methods incorporate a whole range of environmentally friendly products and systems. Green buildings reduce not only greenhouse gas emissions, but also many other forms of pollution and waste.

def•i•ni•tion

A **green building** is an energy- and resource-efficient structure. The term may also refer to construction methods that are energy- and resource-efficient.

The nonprofit U.S. Green Building Council offers a special program to help improve energy efficiency in a wide range of buildings. Its Leadership in Energy Efficiency and Design (LEED) Green Building rating system considers many factors; buildings that meet these rigorous standards earn the Green Building designation. Even their plaque is environmentally friendly, made of recycled glass that's sandblasted instead of chemically etched.

For information about how to retrofit an existing building or incorporate ideas in a new project to earn this certification, visit www.usgbc.org.

Improving the way energy is used within a building can start at the very top of the structure. *Green roofs* use soil or another growing medium plus carefully selected plants to …

- Moderate temperature extremes on the roof.
- Help moderate temperatures within the building below.
- Absorb and store carbon from the atmosphere.
- Absorb and filter rainwater.

def•i•ni•tion

A **green roof** consists of plants, soil, or other growing medium, and a moisture- and root-barrier layer that all work together to conserve energy use within the structure below.

The insulating properties of green roofs mean that less heat from sunlight is transferred to the spaces immediately below the roof. That makes them especially appealing for large, low buildings with surface areas that are measured in acres instead of square feet. Shopping malls, manufacturing plants, warehouses, distribution centers, and other typically sprawling structures are excellent locations for green roofs.

Green roofs are energy efficient in three more ways. They can help make rooftop solar panels operate more efficiently. At extremely high temperatures, solar panels don't function well, so a green roof may be a big improvement over conventional roofing materials that tend to trap and concentrate heat. They also keep heat from radiating out and creating the so-called "heat island" effect, which we'll discuss in Chapter 7. Also, the interior energy use savings provided by a green roof typically occur just when demand for electricity is at its peak, on very hot sunny summer afternoons. That can mean big savings in carbon emissions from power plants.

 Carbon Caution

> Installing a green roof is a job for professionals, due to weight load limits of the supporting structure underneath and the need for plants, soils, and drainage systems specially suited to this unusual growing space.

Reducing the Carbon Impacts of Business Travel

Traveling is an important part of many jobs. Service technicians, delivery personnel, and salespeople may spend their entire workday out on the road. Many jobs often require meetings with co-workers in other locations, as well as travel to distant cities for corporate events, trade shows, and conferences. Let's take a quick look at some ideas that can reduce your carbon footprint whenever travel is part of your job.

Carbon Impact

Use teleconferencing, videoconferencing, and other technology to give an "in the same room" feel to meetings whenever you don't really have to be in the same physical space with the other people.

To save on fuel, service technicians, delivery drivers, and fleet managers can …

- Use computer programs to map the shortest route between stops.
- When time permits, schedule stops in one area on one day, then in another area on another day, to prevent backtracking.
- Make sure vehicles are maintained properly for peak fuel economy.
- Investigate alternative fuels for the fleet, to reduce carbon emissions.

Businesses can encourage carpooling among employees by offering special parking places as a perk. Many communities already have matching services for commuters who want to share a ride—you can find them on the Internet with your favorite search engine by typing in the name of your town followed by the phrase "vanpooling" or "carpooling."

Air travel presents some interesting challenges to the carbon-conscious traveler. Most discussions of air travel and greenhouse gas emissions focus on two bits of advice—keep air travel to the absolute minimum and then, when you must fly, purchase carbon offsets (explained in Chapter 10) to compensate for the effects of burning jet fuel.

 Carbon Extras

Airplanes are a form of mass transit, and that means that they are most fuel efficient when they are full of passengers; every empty seat in an aircraft means a lost opportunity for that flight's carbon footprint to be shared among the maximum number of people.

But other approaches are worth exploring. The ideal plane trip is a direct flight from your base to your destination; every time you land and take off, you increase the amount of jet fuel used and the size of your carbon footprint. When you must travel by air, choose a route with legs that move you closer to your destination, not farther away. This isn't as obvious as it seems. Recently a travel agent suggested I fly a few hundred miles north from my base then make a connection there to my destination, a city far to the south of my base. When I checked some other options, I found I could take a much more direct route (through a city farther south and east), saving plenty of jet fuel because the total distance traveled was shorter.

Not all jet fuel is used in the air—a lot of it is wasted on the ground as jets queue up for take-off or wait in remote spots to taxi in to the terminal. Adjusting your flight times to avoid peak travel times may mean that you spend less time wasting fuel in a jet that's on the ground. Avoiding peak air travel times (whether that's a certain time of day, day of the week, or season) may also mean less time circling in a holding pattern waiting for space to open up on the ground.

Carbon Impact

Improvements in scheduling at major airport hubs could improve traffic flow in the sky and on the tarmac—and cut down on wasted jet fuel.

When an overnight stay at a hotel is part of your job, you can reduce your carbon impact by …

- Turning off lights and the TV when not in the room.
- Adjusting the thermostat.
- Using less water.
- Reusing linens and towels when practical.

Many hotels and motels now have energy-wise programs in place. Look for informative cards in your room that explain guidelines for linen replacement, recommended thermostat settings, and other helpful hints. If you don't find any energy saving info, let the management know about your concerns.

The Least You Need to Know

- Educational programs to improve energy knowledge and awareness among students and teachers play an important role in improving energy efficiency in school buildings—and can reduce carbon footprints for everyone in a community.

- Certified green-building programs and new applications of technology help reduce greenhouse gas emissions for schools, offices, manufacturing plants, and heavy industry.

- Green roofs containing plants and special soils reduce energy use inside the structures.

- Careful planning of routes and timing can help you reduce the carbon impact of ground and air travel for business purposes.

Chapter 5

Carbon from Here to There

In This Chapter

- Comparing motor fuels and fuel systems
- Figuring your miles per gallon
- Carbon impact of public ground transportation
- Motorcycles, scooters, bicycles
- Matching transportation needs to available options

Every time you travel from one place to another, your carbon footprint increases or decreases. Whether your journey is a short visit to the neighborhood grocery store, a family vacation to another state, or a long-distance business trip, how you get from here to there and back again has an impact on greenhouse gas emissions.

In this chapter, we take a look at your various transportation options. I'll give you some handy reference tools and pointers on how to compare choices. And I'll include practical tips so if you decide to try something new, you'll feel comfortable right away.

Comparing Transportation Choices and Motor Fuels

For each trip, no matter the distance, you'll want to consider these factors:

- What kind of fuel does this vehicle use?
- What mileage figures are typical for this fuel in this kind of vehicle?
- Am I traveling alone, or will others be sharing this choice with me?

You don't have to work out an elaborate chart or even have a pencil in hand—these are just some thinking points to keep in mind as we explore all the choices.

Two Ways to Go

Two kinds of internal combustion engines are common. Gasoline engines use an electric spark to get things going. Diesel engines use heat from air compression to get things going. The kind of engine in a vehicle, whether it's a passenger car, a light truck, or heavy equipment, depends on many factors. Generally, diesel engines are the better, more efficient choice for heavier work, so diesel engines are more common in heavy-duty trucks; equipment for farming, mining, and construction; and locomotives.

Throughout the twentieth century, most vehicles were designed to operate with only one kind of fuel. One model would use only gasoline made from petroleum, another only diesel made from petroleum. Today's choices still include many vehicles that use conventional fuels such as gasoline or diesel from petroleum. But in the twenty-first century, there are a lot more fuel choices, and some vehicles can switch from a certain kind of fuel to another.

Alternative Fuels

Let's take a look at the alternatives. *Biofuels*, blends that include varying amounts of petroleum and plant materials, are the most readily available alternative fuels. One of the most common combines gasoline with *ethanol*, sometimes called ethyl alcohol or grain alcohol, which can be made from the starches in grains such as corn, or from the cellulose in grasses or certain kinds of trees. No matter the origin, the chemical properties of ethanol are the same. These are the two most common ethanol blends:

- *E10*, a blend that is 90 percent gasoline and 10 percent ethanol
- *E85*, a blend that is 15 percent gasoline and 85 percent ethanol

E10 is so commonly available today that you may not even realize you're using it. Next time you're at the gas station, look on the front or side of the pump. You may see a small notice stating the ethanol content of the fuel as a percentage that may go as high as 15.

Using ethanol instead of gasoline reduces greenhouses gases. Ethanol made from corn can reduce greenhouse gases by about 25 percent when compared to gasoline. Ethanol made from *cellulosic plant materials* may be able to reduce greenhouse gases up to 100 percent when compared to ordinary gasoline.

def•i•ni•tion

Cellulosic plant materials are the tough fibers such as stems and leaves of plants. Corn stubble (the leftovers after the ears have been harvested) and fast-growing switchgrass could become an important resource for biofuels in America. In other parts of the world, tough plant fibers such as sugar cane are being used as a source for biofuel.

If you're thinking about replacing your current vehicle with one that uses a different kind of fuel, check to see if you'll be able to buy that fuel in your area. One of the not-quite-solved problems with biofuels is that because of the chemical properties of their ingredients, they require a different distribution network than conventional fuels.

Biodiesel is the other most common biofuel. Biodiesel is usually made from soybeans, but it can also be made from certain animal fats. Most biofuels are made from scratch, but it is possible to take used vegetable oil, such as that left over from frying foods at restaurants, and use it as a fuel in specially modified vehicles. Some people find that the vehicles do produce an unusual odor when powered by this recycled fuel.

Carbon Extras

To locate biodiesel anywhere in the United States, visit the National Biodiesel Board's website at www.biodiesel.org.

A comprehensive fuel locator service for all kinds of alternative fuels is available at this U.S. Department of Energy website: www.eere.energy.gov/afdc/fuels/ stations_locator.html.

Three main kinds of biodiesel are available:

- *B2*, a blend of 2 percent biodiesel and 98 percent petroleum diesel
- *B20*, a blend of 20 percent biodiesel and 80 percent petroleum diesel
- *B100*, a plant-based fuel that is a complete substitute for petroleum diesel

Using a biodiesel blend such as B20 can reduce carbon dioxide emissions by about 15 percent. Using B100 instead of petroleum based diesel can reduce carbon dioxide emissions by up to 75 percent. Each biofuel has a different amount of carbon within it. During combustion (burning the fuel in the engine) the fuel combines with oxygen from the

air. The waste products include carbon dioxide.
The amounts of carbon dioxide released after combustion differ depending on the fuel.

Carbon Caution

Before changing the kind of fuel you
use in your vehicle, be sure to check the
owner's manual. For example, the corrosive
properties of ethanol may damage internal parts in the fuel system of some older
vehicles or void certain portions of your
vehicle's warranty.

Ethanol and biodiesel reduce greenhouse gases
because the crops capture and store carbon dioxide
while they are growing, then these are released
back into the air where they can be reabsorbed
during the next growing season, which is part of
the natural carbon cycle.

Innovative Fuel Systems

Reducing greenhouse gas emissions by changing
fuel types is not just a matter of substituting one
kind of fuel for another. In order to use new fuels
it is often necessary to make changes to a vehicle's
entire fuel system. Some innovative options are
more readily available than others, and some are
only in the early stages of development.

There are important differences among the most
talked about new fuel systems:

- A **flex fuel vehicle (FFV)** with an internal combustion engine can be operated with conventional petroleum gasoline, E10, or E85, depending on what the driver puts into the tank.

- A **hybrid electric vehicle (HEV)** uses a conventional internal combustion engine in certain situations but automatically switches to electric power from special batteries, and vice versa, depending on the driving conditions; the driver doesn't control the change.

- A **plug-in electric vehicle (PEV)** uses electricity alone. It can be plugged in to an electrical outlet to obtain electricity from the grid, store that electricity in batteries, then be unplugged and driven using the stored energy in the batteries.

- A **plug-in hybrid electric vehicle (PHEV)** uses a combination of a conventional internal combustion engine with batteries that get their initial charge from the grid.

Many American car dealerships already offer flex fuel vehicles for sale. Several manufacturers offer hybrid electric vehicles, but the technology is not yet well developed enough for the two plug-in types of electric vehicles to be widely available.

Most newer vehicles can use either petroleum gasoline or ethanol blends with up to 15 percent ethanol without any special modifications. There is flexibility built-in. Check your owner's manual for information.

A true flex fuel vehicle is specially constructed to operate with a much higher portion of ethanol in the liquid fuel, using E85, and bears the words "flex fuel" or the acronym "FFV" on the outside.

> **Carbon Caution**
>
> Gasoline engines can be converted to become diesel engines. Existing diesel engines can be converted to accept straight vegetable oil as a fuel. When looking for these types of after-market kits, investigate the manufacturer's claims carefully, and check your vehicle's owner's manual for warnings before you make a kit purchase. Many people who've tried these kits belong to local clubs or online discussion groups. Find out as much as you can before you try this new technology. And be sure you can find your new fuel when and where you'll be driving your modified vehicle.

Each of these fuel systems could reduce greenhouse gas emissions by significant amounts, but there are still many technical and practical issues to be resolved. Later in this chapter, we'll take a look at the plusses and minuses of the various kinds of electric vehicles.

There's another alternative fuel that doesn't get very much publicity, probably because the technology's been around for more than 60 years—and the source is another fossil fuel. Many vehicles can be powered with liquid petroleum gas (LP gas, LPG, or propane). Some vehicles are designed to operate exclusively on LPG, and some have a dual fuel system that can use either conventional petroleum gasoline or LPG.

Any gasoline powered vehicle can be converted to use propane. With lower carbon monoxide emissions than gasoline, LPG is a popular choice in enclosed spaces such as warehouses. In the United States, there about 350,000 vehicles equipped with LP gas fuel systems; worldwide more than 9 million vehicles use this fuel.

LPG for vehicles, also known as *autogas* (all one word), can reduce a wide variety of emissions. Using it may help reduce your carbon footprint because using LPG instead of conventional petroleum gasoline can mean 40 percent fewer hydrocarbons released. Also LPG is thought to have 50 percent fewer potential ozone-forming emissions than conventional gasoline.

Another greenhouse gas reducing fuel system still in the early stages of development would use *hydrogen fuel cells* to propel the vehicle. A hydrogen fuel cell is rather like a battery and engine all in one. In a fuel cell the energy released when liquid hydrogen and liquid oxygen react is converted into a continuous supply of electricity. We'll take a close look at that innovation in the chapter on carbon initiatives.

Figuring Your Gas Mileage

There's a catch to carbon dioxide emissions data comparisons. Not all fuels are alike in how much work they can perform. A gallon of unleaded gas performs a different amount of work than a gallon of diesel fuel, and the amount of miles you can travel while using one gallon of fuel (whatever the type) varies, depending on the size and shape of the vehicle, road conditions, and a whole lot of other factors.

Three related factors contribute to how well-suited a fuel is to a particular job:

- **Fuel economy** describes the number of miles traveled per gallon of gas or other fuel. Another way to say this is how much fuel it takes to move a vehicle a certain distance. Most people say gas mileage when they mean fuel economy.

- **Fuel efficiency** describes how much energy a fuel produces. An efficient fuel produces the most work; an inefficient fuel does not do as good a job.

- **Fuel consumption** refers to all the fuels used for any kind of work. A hybrid electric vehicle consumes two kinds of fuel.

The stickers in the windows of vehicles at car dealerships include an EPA estimate of miles per gallon (fuel economy), one for city and one for highway.

Although many new vehicles include an option for an onboard computer that figures and displays the fuel consumption in terms of miles per gallon right there on the instrument panel as you drive, most cars on the road today don't have this feature.

You can figure your own gas mileage by keeping this handy chart nearby.

Fuel Economy Worksheet

Trip one to gas station (fill tank completely):
Odometer reading   (A)_____

Trip two to gas station (fill tank again):
Odometer reading   (B)_____

Miles driven: (B) – (A) = (C)_____

Gallons to refill tank   (D)_____

(C) ÷ (D) = (E) miles per gallon, your fuel economy

Fuel Economy Worksheet Sample

Trip one to gas station (fill tank completely):
Odometer reading   (A) 36,129.7

Trip two to gas station (fill tank again):
Odometer reading   (B) 36,516.8

Miles driven: (B) – (A) = (C) 387.1

Gallons to refill tank   (D) 19.249

(C) 387.1 ÷ (D) 19.249 = (E) 20.11 miles per gallon

You can bet your real-world mileage is different from your vehicle's EPA sticker. That's because most of us drive a combination: some city stop-and-go, some highway cruising. Whether you're using the air-conditioning, have the windows rolled down, and lots of other things can also change your fuel economy. I like to do the mileage check three times in a row and then take an average of the three tankfuls. I do this twice a year, once in mid-winter and once in mid-summer. The results are different.

Boost Mileage, Drop Consumption

Increase your fuel economy and decrease your carbon footprint with these ideas:

- Get rid of extra weight in your car. No need to carry around your golf clubs in the trunk all the time if you only play once a week.

- Keep tires properly inflated to reduce drag and improve efficiency.

- Use four-wheel drive only when you really, really need it for performance.

- Have more than one vehicle in your household? Use the one with the best gas mileage and most appropriate passenger/cargo area for as many trips as possible.

- Travel at a steady speed when you can. Accelerating uses extra fuel.

- Avoid jackrabbit starts and abrupt stops.

- Slow down. Most engines operate most efficiently at lower speeds.

- Plan trips to avoid known rush hour traffic jam areas. Idling the engine in traffic wastes fuel.

Carbon Impact

If you're going to be stopped out of traffic (such as at the curb or in a parking lot) for more than one minute, it saves fuel to turn off your engine until you're ready to move again. Save energy by skipping idling at a drive-thru—just park and walk inside.

Even if you keep your conventional gas vehicle well maintained to get the best fuel economy, there's another problem. Gas engines are not very fuel efficient. Gas engines in cars convert only about 20 percent of the potential energy in the gas into usable energy for mechanical work. All the heat that goes to the radiator or out the tailpipe is energy that did not propel the vehicle—it's wasted.

So four fifths of the carbon dioxide going into the atmosphere from a gas-powered vehicle didn't really help you get from here to there. Would you buy a chocolate malt, drink one fifth of it, and pour the other four fifths down the kitchen sink? I sure wouldn't, yet that's the kind of waste that happens every time you or I drive a gas-powered vehicle. That's why so many people are trying to figure out better ways to power vehicles.

Share the Ride

Another way to reduce your carbon footprint is to carpool. That way, you and your fellow riders can divide the carbon emissions from just one vehicle among yourselves.

Many cities have special traffic lanes (often called the diamond lanes) for high-occupancy vehicles (HOVs) for carpoolers and people using mass transit such as vans and buses. HOV lanes cut down on traffic congestion, so they reduce carbon footprints in more than one way. HOVs and HOV lanes are part of a program to make it more appealing for individuals to share the ride with others.

Restrictions vary from community to community, but in general, HOV lanes are only for passenger vehicles and trucks with two or more occupants, and vans and buses. Some HOV lanes also permit flex fuel vehicles with special stickers or license plates obtained in advance; some HOV lanes have special hours of operation. The symbol for an HOV lane is a diamond.

A Closer Look at Electric Vehicles

Many experts believe it will be easier to control carbon emissions at one central location, such as a power plant, than on millions of individual vehicles. That's one of the reasons electric vehicles are getting so much attention. Electric cars can help reduce carbon footprints two ways:

- Reduce tailpipe emissions
- Increase fuel efficiency

Most of the electric cars available today are called hybrid electric vehicles (HEVs) because they run on a combination of gasoline and electricity. Some engineers are also working on developing a hybrid diesel-electric vehicle for road use; that fuel combination is already in use in locomotives.

Some auto manufacturers are also working on plug-in electric cars (PEVs) that operate exclusively with electricity; the user would plug the car into an outlet to charge the vehicle's batteries, unplug it, and then drive away. It's a lot like a golf cart, but it still needs a lot of technical improvements.

Many scientists (and amateur inventors, too!) continue to experiment with all-electric, solar-powered cars, but practical technology has not yet been developed. So let's take a look at hybrid electric cars and how they can reduce carbon footprints.

A hybrid electric vehicle's power comes from two sources, the gasoline for certain situations and electricity stored in batteries for other situations. The car is constructed in such a way that the driver doesn't need to make any decisions because the car automatically shifts between power sources as needed.

One of the main difficulties so far with making HEVs practical concerns the difference in weight using these two kinds of fuel. With today's technology, about 1,000 pounds of batteries are needed to store the same amount of energy available from a 7-pound gallon of gas. That means that an HEV may have a different weight than a typical car powered by an internal combustion engine alone. And those large, heavy batteries take up space that would be available for passengers or cargo.

There's a very important reason to use electricity instead of gas to power cars. Electricity can be produced from sources such as hydro, wind, and solar that emit little or no greenhouse gases.

Electric cars, whether they're hybrids or totally electric, do have another advantage. Electric batteries are much more efficient than conventional gas-powered engines; some scientists say the ideal electric car could have a 72 percent fuel efficiency rating. That's a really big difference from the 20 percent fuel efficiency rating of a typical gas-powered vehicle.

 Carbon Extras

One of the interesting things about HEVs is that the motor can do two things: in certain situations, it draws power from the batteries, and in other situations, it can produce enough surplus power from the conventional fuel to recharge the batteries. That means that the liquid fuel is providing energy twice, once when the fuel is combusted and again when the energy it produced and stored temporarily in the batteries is eventually used.

A further inspiration for work on electric cars is that new technology being tested today could someday soon dramatically improve the fuel efficiency and reduce greenhouse gas emissions at fossil fuel power plants. That would make using an electric vehicle an even better choice for more people in the future than it already is today.

Other kinds of nonpetroleum fuel cars are being developed right now. Although some are being tested in real-life driving situations, they are not yet readily available, so we'll talk more about that in Chapter 9 on carbon initiatives. That's where I describe the Hydrogen Fuel Initiative.

Mass Transit and Micro Transit

Just how important are buses, subway cars, passenger rail cars, and other mass transit options in reducing greenhouse gas emissions?

Mass-transit vehicles such as buses don't have very impressive fuel economy numbers. But they do have good fuel efficiency when they are transporting large numbers of people. Trains and buses use about 3 percent of the total energy used for transportation in the United States.

Increasing the number of people who go from here to there in just one mass-transit vehicle reduces carbon footprints in several ways. If the fuel efficiency and fuel economy of the mass-transit vehicle are shared among enough passengers, the net effect can be much lower than for a lot of individual cars. And mass transit can help lower traffic congestion, which means the people who are driving individual vehicles waste less energy while stuck in traffic jams.

During 2005, riders took 9.8 billion trips—totaling almost 50 billion (with a *b*) miles—on America's 150,000 public transportation vehicles.

In a year's time, all those trips:

- Saved 1.4 billion gallons of gas
- Saved 34 supertanker trips to bring gas to the United States from the Middle East
- Saved 140,000 tanker truck trips on American streets delivering fuel to local stations

Another mass transit option for longer journeys is the airplane. But since planes are often involved in leisure activities as well as business trips, I discussed the carbon footprint impacts of aviation in Chapter 4.

At the opposite end of the transportation spectrum are various mini forms of transportation. I think of them as micro transportation options.

Many manufacturers are now offering extremely lightweight mini cars that are about half the size of a typical sedan. These cars often feature tiny engines with fewer cylinders, yet they can still accelerate to highway speeds. Often they have room for only a driver and one passenger, and very little cargo space. Yet with fuel efficiency estimates of between 40 and 70 miles per gallon, they could be an important step toward reducing greenhouse gases.

Motorcycles and scooters with internal combustion engines have excellent fuel economy, often as high as 70 miles per gallon. Mopeds combine human power (feet moving the pedals) with a small internal combustion engine to achieve even better fuel economy. Bicycles are completely human powered and use no other fuel, so they have no primary carbon footprint.

These individual two-wheeler choices are excellent ways to reduce one's carbon footprint, but they come with drawbacks. There's little or no cargo area, weather conditions may make them impractical, and traffic laws may inhibit their use in certain circumstances.

Making Better Choices

In this chapter, we've looked at a lot of transportation choices, some that are readily available and some that aren't yet commonly in use.

One of the best ways to reduce your carbon footprint consistently may be to begin breaking each journey into segments and then using the most carbon-friendly choice for each portion.

Here's an example. You might drive your personal vehicle to a meeting spot and pick up one or two other people and carpool to a point close to your common destination area. Once there, each of you might finish the journey by choosing different options. One person might ride the bus for another mile, another might use a bicycle stored nearby to go several blocks, and another person might simply walk the rest of the way. Later in the day, you all reverse the process.

Many localities are encouraging just this sort
of thing with programs such as free or low-cost
loaner bikes, special parking areas for park-and-
ride mass transit services, and free commuter
matching services for folks who want to join a vanpooling or carpooling effort.

The Least You Need to Know

- Maintaining your current vehicle in peak
  condition and driving it carefully can
  improve your fuel economy and reduce your
  carbon footprint.

- Choosing different fuels and/or fuel systems
  for personal vehicles can reduce greenhouse
  gas emissions right now.

- In the future, reductions of greenhouse
  gas emissions at power plants could make
  vehicles powered in whole or in part by
  electricity an easy way to reduce one's carbon footprint.

- Matching the most energy-efficient form
  of transportation that produces the fewest
  greenhouse gases and suits your needs with
  each segment of a journey is a good way to
  consistently reduce your carbon footprint.

Chapter 6

Lights, Computers, and Communications Networks

In This Chapter

- How different lights use energy
- Preventing light waste
- Matching energy use to real needs
- Hidden energy use

In our industrialized, high-tech society, we use lights and electronic technology around the clock, indoors and out. We light up our rooms even when daylight could do the job for us. We light up the night sky. We use light for information and fun, advertising, even art.

We use lights in our homes and workplaces, in schools and shopping malls. We depend on lighting for safety on our streets and highways. We have computers on our desktops, cellphones in our pockets, televisions everywhere—and all of these electronics use energy, too.

We enjoy blinking displays in Las Vegas and Times Square, on the jumbo screens at athletic events, and on our houses and Christmas trees at holiday time. Colorful signs in our own neighborhoods tell us what movie's playing, the time and temperature as we drive up to the bank, which restaurant's open—the list and the lights go on and on.

In this chapter, we take a look at the interwoven areas of lighting choices, computer technology, communications networks, and entertainment to see how these activities contribute to the size of your carbon footprint.

Two Common Lighting Choices

The two most common kinds of lights are …

- Incandescent.
- Fluorescent.

Both types have important differences. In an incandescent bulb, electricity passes through a metal wire called a filament. Some of the energy causes the filament to glow, and the wasted energy escapes as heat.

In a fluorescent bulb, as the electricity passes through special gases, it causes the coating on the inside of the glass bulb to glow. Very little energy is wasted as heat. Fluorescent lights are much more energy efficient than incandescent lights.

Light is measured in lumens. Each watt of electricity going to an incandescent bulb produces about 15 lumens of light, while each watt of electricity going to a fluorescent bulb produces between 50 and 100 lumens of light. Fluorescent lights also last up to eight times as long as incandescent bulbs.

Carbon Extras

Electricians and other professionals use the word *lamp* when you and I use the word *bulb* or *light bulb*. In this book, we use *bulb* more often because, to most people, a *lamp* is something pretty sitting on a table.

Those conventional long, straight fluorescent lights have been an energy-efficient choice for lighting large areas such as schools and businesses for years. However, these long tubes need special equipment (starters, bases, and such) to work and special fixtures to hold them. Their use in homes has been limited to a few areas due to their size, as well as problems with color balance and complaints about flickering and buzzing.

The *compact fluorescent light bulb* (CFL) changed that. These all-in-one bulbs feature tiny gas-filled tubes connected to bases that will fit anywhere a conventional incandescent bulb can. The quality of light is excellent in the latest generation of CFLs.

def•i•ni•tion

> A **compact fluorescent light bulb (CFL)** is a small bulb with bent or spiraled gas-filled tubes in a single base that can be substituted in most sockets for a conventional incandescent bulb.

So CFLs are recommended today to replace incandescent bulbs because they require much less energy to provide equivalent light and they last much longer. What's not to like?

Well, there are still some problems with CFLs. They do not work well in extremely cold conditions. They also do not work well in areas that are subject to a lot of vibration, such as in ceiling fans. And using a CFL where you want to be able to dim the light requires special equipment. (Consult a lighting supplier for proper installation.)

But the biggest drawback could be what's inside them. Fluorescent bulbs contain mercury, a known environmental hazard. A typical CFL contains only 5 milligrams of mercury, but that's enough to pose a safety hazard if the bulb breaks or when it finally wears out. Broken or used fluorescent bulbs of any kind, including CFLs, must be disposed of properly, something most consumers don't realize or know how to accomplish.

For complete details on the proper disposal of CFLs, visit www.energystar.gov/ia/partners/promotions/change_light/downloads/Fact_Sheet_Mercury.pdf.

Carbon Caution

Use extreme caution when cleaning up a broken fluorescent bulb. Do not vacuum up the particles. Sweep gently and use a damp cloth to pick up the pieces; then seal everything in double plastic and take it to a hazardous material recycling center.

Turning off lights when you don't need them, even if it is just for five minutes, is sound advice when the light is an incandescent bulb. But due to different operating principles, any kind of fluorescent light should be turned off only if the period of nonuse is greater than 15 minutes. Turning a fluorescent light off and then on again within 15 minutes can shorten its useful life.

Other Lighting Innovations

Many other lighting choices are making it easier to match the right light source to the situation and to reduce energy use.

Two other classes of lighting use bulbs. *High-intensity discharge (HID)* lamps (bulbs) include mercury vapor, metal halide, and high-pressure sodium. All offer these advantages:

- Ability to spread a lot of light from one source over a large area
- Excellent energy efficiency
- Very long life

def•i•ni•tion

High-intensity discharge (HID) lamps (bulbs) contain gases and metals that provide very strong light very efficiently.

Typical uses for HIDs include large outdoor areas such as parking lots and highway interchanges, as well as outdoor plazas and recreational areas. Improvements in the metal halide style of HIDs increase their suitability for some indoor uses, even in homes.

Extremely energy-efficient *low-pressure sodium lamps* (bulbs) are also useful outdoors, but their large size (up to 4 feet long) and odd color effects (things look different than in natural daylight) limit their appeal. Low-pressure sodium lights are commonly found above sidewalks, roads, and parking lots.

def•i•ni•tion

A **low-pressure sodium lamp** includes gases that vaporize to create a distinctive yellowish-orange light; it works on a completely different principle than a high-pressure sodium HID light.

Low-pressure sodium lights are the most energy-efficient, readily available lights on the market, capable of producing as many as 180 lumens from a single watt of electricity. One low-pressure sodium light can work for up to 18,000 hours. These are excellent energy savers!

But some of the most useful new light sources don't use a bulb. This new category is called *solid-state lighting*. Solid-state lighting produces light without heat as a by-product. There are two kinds:

- *Light-emitting diodes* (*LEDs*)
- Organic light-emitting diodes (OLEDs)

def•i•ni•tion

> Energy-efficient **light-emitting diodes** (**LEDs**) produce light from electric current passing through tiny semiconductors instead of using gases or other extra materials.

An LED contains a tiny semiconductor (a bit of special material) that emits light when electricity goes through it. No additional wires or gases are needed, nor any kind of glass covering like in a typical light bulb. LEDs function with only a thin sheet of plastic covering them. They don't emit heat, either, so they are very energy efficient. They also have a long working life.

You're already very familiar with the style of LED that makes all those little red or green numbers on microwaves, VCRs, and other devices throughout your home. That was pretty much their limit until inventors got busy making improvements to overcome a few LED drawbacks.

LEDs have two disadvantages—a single LED doesn't produce many lumens, and the light from an LED can't be directed very well. LEDs work best when viewed straight on, and that's how most are used today. Grouped together in tight arrays, they've become a reliable, energy-saving part of traffic signals. Cities that replace conventional incandescent bulbs with LEDs in traffic signals and highway signs save energy and lower everyone's carbon footprint. Even newer applications use LEDs within highway signs. But you're probably most familiar with LEDs as a part of those jumbo TV screens at sporting events and in Times Square in New York.

Dramatic improvements in white LEDs mean they're becoming useful for lighting small indoor areas, such as bookshelves, and small outdoor areas around homes, such as steps and paths. Early LEDs glowed only in red or green, but technological advances make many more colors available today. That's why LEDs are now turning up as an energy-efficient improvement over neon signs. Encased in easily shaped thin plastic strips and available in many colors, LED strips can spell words, create picture designs, and outline building edges with ease. Although the initial cost is often still higher than with neon, the operating costs are much lower because of the huge energy savings.

Organic light-emitting diodes (OLEDs) work by passing electrical current through very thin films of various kinds of organic and inorganic materials; as energy moves from layer to layer, light is emitted. OLEDs are not light bulbs, but rather a form of display that uses light.

def•i•ni•tion

Organic light-emitting diodes (OLEDs) produce light as electrical energy moves through layers of materials.

The most widespread use of OLEDs so far is in small devices such as cellphones and digital cameras. Since OLEDs emit light themselves, they are a good replacement for older kinds of display-screen technology that required backlighting. That saves energy during operation. Also OLEDs are not as heavy or bulky as those older methods, which can save on the amount of energy used to transport them to where they'll be used.

A Little Less Light

Finding the best light for the job involves more than matching up the number of lumens, the color of the light, and the amount of power needed to produce the light.

One of the biggest energy wasters in lighting isn't the wrong bulb—it's the wrong approach. Light that doesn't go exactly where it's needed uses up valuable resources. Properly directed light saves energy by putting the light only where it's needed. Swivel mountings, shields, and other physical changes can mean fewer light fixtures and fewer lumens are needed to achieve the same results.

The easiest way to start recognizing light waste is when you're outdoors at night. Light and the energy to provide it are being wasted when …

- Light goes beyond its intended location, such as up into the sky.
- Light glares off nearby surfaces, interfering with your vision.
- Lights are spaced incorrectly, causing visual clutter and confusion, and annoying alternating bright and dim spots.
- Lights are on at improper times, such as when an area is empty.

At home, you can save energy on outdoor lighting by …

- Repositioning, reducing, or removing strictly decorative landscape lighting.
- Using timers, daylight sensors, and motion detectors to turn lights off and on only when they're needed.
- Using smaller output lights that direct light only at safety locations, such as stairways and path edges, instead of over an entire area.

Once you get a little practice looking for these situations outdoors, it's easy to use the same skills indoors to check for wasted light. Do you need a table lamp for reading and overhead lights at the same time? Do you need a strip of six 100-watt bulbs above the bathroom mirror just to brush your teeth, or will a 15-watt nightlight, light from an adjacent room, or natural light through a window or skylight do just as well?

Carbon Impact

Free-standing solar-powered lights may be a good choice to avoid the use of electricity entirely; try a few at first to see if your local weather provides enough sunshine to keep them working reliably.

Energy for Computers, Cellphones, and Other Surprises

Noticing and correcting wasted energy from lights is easy—you can see it. But figuring out what's going on energy-wise with computers and other common everyday devices isn't so obvious. You can reduce your carbon footprint with some simple steps once you understand what's going on behind the scenes.

A desktop computer saves a great deal of energy by using less paper, but it does use electricity. Desktop computer systems are becoming more energy efficient through improved designs. One of the most significant changes is the invention of "sleep" mode for monitors and processors. When there's no keyboard or mouse activity for a certain time, the system shuts most of itself down and goes into a less-energy-consuming stand-by mode.

Carbon Impact

Sleep modes for computers still consume energy; for even more energy savings, turn your system completely off during the hours you don't need it. Plugging several devices into a power strip can make it easier to save energy by shutting everything down with the flick of one switch. A separate power strip for all your battery rechargers is a good idea, too.

E-mail, faxes, and even teleconferencing all save energy, too, because they often mean less travel for humans and packages. But when you're working online, you're using more than the computer right in front of you. The servers that are the backbone of local computer and communications networks and the Internet require large amounts of energy. Your carbon footprint extends farther than you realize.

Rooms filled with these computers must be air-conditioned 24 hours a day, year-round, to keep sensitive electronics at the correct temperature for proper operation. That's because the computers themselves produce so much waste heat. Improvements in the way the internal parts of these machines are arranged, combined with better arrangement of equipment within a room to improve air flow, could result in substantial energy savings.

Engineers and energy-efficiency experts are trying these ideas to reduce the carbon footprint of large computer installations:

- Using filtered outside air for cooling, when appropriate
- Rearranging equipment for better air flow
- Using rooftop solar panels and biomass to generate electricity locally

Carbon Impact

Less than half the energy used in large data centers actually powers the computers—the rest goes to air conditioning to keep the equipment cool.

Telephone systems use electricity in ways you might not think about when you're talking. Old-fashioned land lines use a very tiny amount of electricity to convey signals and voices. Using a wireless system that taps into that landline uses additional electricity in two ways, first to power the base for the radio waves, and again each time you recharge the mobile handset.

Cellphone networks use electricity in even more ways. Recharging the batteries of the handset requires electricity. Cellphone towers are really mini radio transmitters that also use electricity to operate. Even when you're not talking or text-messaging or otherwise actively using your cellphone, the phone is sending brief intermittent signals to stay in touch with the network.

Using conventional landlines for phone conversations and faxes uses less energy than wireless communications systems. To reduce your carbon footprint, use land-based systems whenever possible, and save wireless communications only for the times they're really necessary.

Carbon Impact

Recharge cellphones, other wireless phones, walkie-talkies, PDAs, and all electronic devices during off-peak times to reduce your carbon footprint; turn them completely off when you don't really need to be in touch with the network.

The Least You Need to Know

- Different kinds of light bulbs and lighting devices use different amounts of energy and produce light of different quality, color, and strength.

- Matching the correct style of light with its purpose and positioning it correctly can prevent waste and save energy.

- Computers, cellphones, and other communications and entertainment networks use more energy than you might realize.

Chapter 7

Carbon Outdoors

In This Chapter

- Urban heat islands
- Landscaping and lawn maintenance
- Outdoor leisure activities
- Team sports and more

Human activities and technology are at the very heart of the discussion of global climate change. What we do at ground level with the land, whether it's a desert in Arizona or a rocky meadow in Maine, and what we do on our lakes and rivers can have a profound effect on the atmosphere high above us.

The United States extends for 3,787,428 square miles. A bit more than 251,000 square miles of that area is water. From the Great Lakes to the Gulf of Mexico, from tiny mountain streams in the Cascade Range to the Savannah River in South Carolina, the American landscape is full of opportunities for outdoor fun. We enjoy boating and fishing and snorkeling, day hiking and overnight camping, skiing and snowmobiling. We've built soccer fields, football stadiums, golf courses, and race tracks for horses and cars.

In this chapter, we explore the carbon impacts of some outdoor situations and activities.

Hot Times in the City and Beyond

Have you noticed that it isn't just our cities that are getting bigger? The things we build in them are bigger, too—taller skyscrapers, more massive highway interchanges with three and four levels of ramps, bigger shopping malls, sprawling parking lots. All that asphalt, concrete, and steel of the built-up human world retains heat. Scientists call it the *urban heat island* effect.

def•i•ni•tion

> An **urban heat island** occurs when the buildings and other structures of a human settlement absorb and retain more heat than the natural areas nearby.

The higher air temperatures of urban heat islands occur for these reasons:

- Concrete, asphalt, and other building materials have different abilities to absorb and release heat than plants, soil, and water.

- Human activities such as operating vehicles and other equipment such as heating and air-conditioning systems contribute waste heat to the air.

- The geometry and placement of buildings and other structures interrupt natural air flow patterns.

Carbon Impact

Summer is the peak time for electricity usage, so reducing heat islands could reduce carbon footprints.

Scientists are not yet sure how significant urban heat islands are in relation to global climate patterns because these islands are essentially local phenomena. A study conducted by the United Nations Intergovernmental Panel on Climate Change reported that although the temperature difference between an urban heat island and its outlying rural area can be as much as 2°C to 6°C on a particular day, all urban heat islands taken together might translate into an increase in warmth in the entire atmosphere of 0.006°C per decade. No one knows yet if that is enough to make a significant difference to climate patterns.

What's certain is that urban heat islands do have an immediate impact on local energy use. During the winter, heating systems may not need to work quite as hard, but that extra heat at other times of the year means that air-conditioning systems have to work harder and longer during the spring, summer, and fall. That sets up an escalating energy-use cycle, as the equipment works harder and adds even more heat to the local air.

Smaller heat islands can occur anywhere there are large expanses of concrete or other paving materials. You can feel this yourself the next time you're in a sunny asphalt parking lot that borders a grassy area. Stand in the parking lot and then lean over with your palm down at about knee height. Feel how hot the air is here over the pavement. Then walk out into the grassy area and hold out your other palm—you can feel that the air here is much cooler.

Here are some ways to reduce the potential for heat buildup in outdoor areas:

- Install green roofs of special plants and soils on the tops of buildings.
- At ground level, break up the heat-trapping surfaces of parking lots with sections of grasses, shrubs, and trees.
- Along sidewalks, use planter boxes and in-ground areas to add vegetation and shade.
- Add mini parks and other green open spaces in urban areas to improve air flow around buildings.

In city and suburban neighborhoods, small buildings and homes can benefit from the proper placement of plants. Deciduous trees (those that lose their leaves in winter) can provide shade in summer, reducing the need for air conditioning, and then allow sunlight to warm things up in winter.

Landscaping and Carbon Impacts

Choosing and maintaining plants for urban environments used to mean simply finding plants whose roots wouldn't heave up the sidewalk or whose limbs wouldn't tangle in overhead power lines.

Today new concepts for landscaping everywhere incorporate several carbon-friendly ideas.

Some New Ideas for Plants

More than 25 years ago, Denver Water copyrighted the term *Xeriscaping* for garden and landscape practices in naturally dry areas. The preferred plants either are extremely drought tolerant or require minimal irrigation; the concept also recommends mulches to retain moisture and control weeds.

def•i•ni•tion

Xeriscaping is a trademarked term for an environmentally friendly form of landscaping that uses minimal watering and requires very little maintenance.

Another movement toward more sensible gardening includes the use of native plants instead of exotic ornamentals. The idea here is that a plant that grew in North America before European settlement is well adapted to the soils and weather of its region and should thrive with little or no human intervention. Using native plants in landscapes can reduce carbon footprints because they …

- Conserve water.
- Require fewer fertilizers.
- Are easy to maintain.

 Carbon Caution

Digging up wild plants is illegal (not to mention harmful) in many locations; buy from reputable native plant suppliers, or share seeds and seedlings from other local gardeners.

The emerald green grass of a typical American lawn may not be the atmospherically friendly carbon sinks they seem at first glance. That's because the typical grass lawn is a monoculture of one grass species that often requires artificial fertilizing, irrigation, chemical pest control, and constant mowing to achieve its uniform appearance. Each of those activities contributes to greenhouse gas emissions. Today many carbon-conscious advisors recommend these steps for a more sustainable lawn:

- Plant a mixed lawn of several grass species.
- Choose very low- and slow-growing grasses that require less mowing.
- Tolerate a few weeds and brown spots.

- Add more native grasses and wildflowers, and then let all the plants grow taller than a typical lawn and mow only a few times during the growing season.

- Increase the space devoted to flower beds, shrubby borders, and trees to decrease the area for a lawn.

Carbon Impact

Groundcovers of vines and other kinds of low-growing plants are a good substitute for high-maintenance grasses.

Equipment for Lawns and Outdoor Maintenance

Until lately, lawnmowers, leaf blowers, and other sorts of small-engine equipment for outdoor situations didn't get the attention that motor vehicles do. They aren't used as frequently, and their gas tanks are small. But using these small engines has a big impact on the atmosphere, due to emissions of both criteria air pollutants (which we'll discuss in Chapter 9) and greenhouse gases.

Replacing gas-powered equipment with electric equipment offers plusses and minuses:

- Electric equipment does not emit criteria air pollutants or greenhouse gases—but the power plant supplying the electricity probably does.

- Using electric equipment on hot summer afternoons may increase peak demand when electricity-generating plants are already working at maximum levels.
- Different cutting widths, as well as differences in horsepower and torque, may make electric mowers less efficient in certain circumstances.

The biggest argument in favor of switching to electricity-powered lawn equipment is the same one advanced for electric cars. It may be easier to control emissions effectively at a single power plant than from thousands of widely dispersed pieces of equipment.

 Carbon Extras

The EPA and members of the power equipment manufacturing industry are developing new emissions controls and standards for small engines that will become effective in 2011.

Whatever kind of mower you use, follow these steps for maximum energy efficiency:

- Keep blades sharp.
- Set blades to their highest level.
- Keep all air filters and moving parts clean.

In many situations, an even better choice may be human-powered tools. Using rakes, brooms, hand clippers, and even push mowers (known in the trade as reel mowers because of the way they operate) eliminates emissions.

Carbon Caution

Know your own strength before buying a reel mower—pushing one across a concrete showroom floor at your local home products store is a lot easier than pushing one through 6-inch-high grasses. Ask for a free in-yard trial period.

A mulching-style mower that chops clippings into tiny pieces is environmentally friendly because …

- Clippings and leaves that remain on the lawn return nutrients to the soil, retain moisture, and may reduce or eliminate the need for chemical fertilizers.

- No bags are needed, no trucks have to pick up those bags of yard waste, and no landfill space is used.

Starting your own compost heap can prevent greenhouse gas emissions by eliminating the need for waste-hauling trucks; adding the finished compost back to your lawn and garden areas improves soil quality without the use of artificial substances.

Outdoor Fun, Games, and Greenhouse Gases

In 2007, cars in the Indianapolis 500 switched from normal gasoline to a blended fuel that's 98 percent ethanol. That's the highest-profile example of energy consciousness appearing in the world of sports.

In this section, we take a look at how energy-saving ideas are affecting a few common outdoor pursuits.

Carbon Impact

Take your ski racks (or bicycle rack, cargo boxes, or any other outside storage items) off your vehicle when you don't need them, to reduce drag and improve gas mileage.

Ski resort operators recently joined together to help reduce greenhouse gas emissions by ...

- Replacing old equipment with more energy-efficient items.
- Installing solar equipment at lodges and lift shacks.
- Adding wind power–generating stations.
- Using energy-efficient building technologies for new construction.

- Offering skiers on the slopes ways to participate in programs that encourage the development and use of renewable fuels. One Colorado ski area offers skiers who arrive together in high-occupancy vehicles prizes as a reward for their energy-saving transportation choice.

Outdoor stadium managers, indoor arena managers, swimming pool and tennis club owners, local health clubs, and even bowling alleys and ice-skating rinks are getting into the action to reduce energy use in their facilities. The same energy review and improvement programs (such as LEED) offered to other kinds of businesses and homeowners are also available for sports facilities. Owners and managers are choosing more energy-efficient lighting and better HVAC equipment, and making improvements to every aspect of their buildings to reduce greenhouse gas emissions. They're monitoring water usage, too.

Carbon Impact

On game day, reduce your carbon footprint by joining other fans in a carpool or by taking advantage of shuttle bus services from outlying meeting points to the stadium.

When golfing, instead of using a battery-powered golf cart for every round, use a pull cart or a human caddy for your clubs as often as you can.

The Golf Course Superintendents Association of America recently began a new program to survey its members on a wide range of environmental issues and practices, including turf management and water usage. Their aim is to collect statistics and then provide meaningful ideas to their members on how to conserve energy and make this outdoor sport more environmentally responsible.

Water sports and activities are an area where alternative fuels may have a big impact on decreasing carbon emissions. Recreational boaters can improve fuel efficiency right away with these steps:

- Keep your engine tuned up for maximum performance.
- Match equipment and supplies to the activity, to cut down on excess weight.
- Keep the hull clean, to reduce drag.
- Use the right propeller and keep its blades smooth.

Improvements in fuel injectors and four-cycle engines for boats may make a new engine a sensible choice to reduce greenhouse gas emissions.

Greening the National Park Service, the name of a 1994 Executive Order, has already meant major improvements in energy efficiency at U.S. parks and historic sites. Park Service employees have been implementing energy-saving practices inside lodges and other buildings, and outdoors with lighting and maintenance equipment. Many also offer park visitors activities that raise energy consciousness, especially through education programs for children. Individual state park systems are also reevaluating their energy use and incorporating energy-saving measures in buildings and outdoor areas.

The Least You Need to Know

- Proper selection and placement of living plants can moderate heat islands caused by buildings and pavements.

- Reducing the carbon impacts of lawn and landscape maintenance can be accomplished through careful plant choices and by substituting electric motors or human power for tools and equipment that consume fossil fuels.

- Emissions controls and new standards for fuel efficiency will make it easier for consumers to compare various kinds of outdoor equipment beginning in 2011.

- Sports facilities and park systems are improving energy efficiency and reducing carbon footprints through many of the same programs available to other businesses and homeowners.

Chapter 8

Carbon in the Food Chain

In This Chapter

- Agribusiness energy use
- Other farm options
- From the field to your plate
- Seasonal variations

Whether they raise animals or plants, U.S. farmers do an amazing job of providing us with a wide variety of wholesome foods every day of the year. And we import common and specialty items from other farmers around the globe. The selection and quality of foods and beverages available in the United States is astounding.

Yet while we're bagging up our purchases, few of us could tell very much about how this bountiful harvest got to the check-out lane. That's because in just over 100 years, the population patterns in the United States have completely reversed. Today only a tiny percentage of Americans lives on or near farms.

In this chapter, we check out some modern farming methods and see how what we eat and drink gets to our tables—and how the choices we make about all our foods impact the carbon cycle.

Energy Use on the Farm

At the beginning of the twentieth century, 98 percent of Americans lived in rural communities and understood from daily contact with animals and plants exactly how food gets from the farm to the table. Today, at the beginning of the twenty-first century, only 2 percent of Americans live and work in rural areas with daily experience in agriculture and animal husbandry.

Agriculture isn't just about food for people. Agriculture includes growing plants to feed animals, as well as growing plants and animals to produce fibers and other ingredients for many nonedible things. Growing plants that can be used in a variety of ways to provide fuel is also an increasingly important role for U.S. farmers. (Biofuels are discussed in Chapter 5.)

Some plants can be grown for more than one eventual use. You'll often see references to *feed*, *food*, and *fuel*. What's the difference? Feed is plant material that is used to nourish livestock. Food is plant or animal material that is used to nourish people. Fuel is plant material that can be used to produce energy.

From small traditional family farms to gigantic corporate holdings, managing energy is now a top priority. It's all part of the movement toward sustainable agriculture. Sustainable agriculture uses resources wisely, maintains a healthy environment, preserves soil and water for the future, and brings long-term economic benefits to farm communities.

Farmers want to be energy efficient because they want to be good stewards of the land and resources. Farmers want to keep their operating costs low to increase their earnings. And today they're also responding to demands from their customers who are interested in reducing greenhouse gases.

One of those demands includes offering more *organic* fruit and vegetable choices, as well as animal products that use no drugs, synthetic chemicals, or hormones.

def•i•ni•tion

Organic items (such as vegetables, fruits, or grains) are grown without the use of artificial fertilizers and pesticides. Organic fertilizers come from other plant or animal materials. Organic pest control is achieved using substances obtained from plants or animals—or, in some cases, living animals such as beneficial insects.

Altogether, these varied American agricultural activities account for about 10 percent of our nation's greenhouse gas emissions. In Chapter 1,

I mentioned that many activities have a primary carbon footprint and a secondary carbon footprint. This is especially easy to see in the agricultural sector.

Carbon Impact

At first glance, choosing organic items seems like an easy way to reduce your carbon footprint because the farmer may have used less energy to grow the item. Many organic items do have a lower primary carbon footprint, but organic items are still a very small portion of the entire food supply. They are often handled separately from conventional produce, which may mean their secondary carbon footprint is equal to or, in some cases, higher than nonorganic items.

Here are some of the energy steps involved in growing corn:

- Fuel for equipment to prepare the soil for planting
- Fuel for equipment to sow the seeds
- Fuel for equipment to spread fertilizer
- Fuel for equipment to control weeds or pests
- Fuel to harvest the crop

Those are all part of the primary carbon footprint of this activity. After the harvest, the corn goes through many additional steps to reach the eventual user. The corn must be transported to a location to be prepared for eventual consumption. That might be a cannery, a frozen food processing plant, or a grain bin. After that, the corn must be packaged in some way and carried to the customer, who could be another farmer raising livestock or a family sitting down to Monday night supper. All these after-harvest activities are part of the corn's secondary carbon footprint.

Most decisions about how to get one thing from here to there are made by evaluating the length of time available, the distance to be traveled, and other factors such as the weight and bulk of the item. The cost in dollars is woven into the decision, too. It wouldn't make any practical or financial sense to transport all the bricks for a house from the kiln to your suburban lot by overnight air-freight! As we'll see, some food transportations choices are not very sensible, either.

 Carbon Extras

The average food travels about 1,500 miles before it reaches the consumer.

In agribusiness, transportation decisions have their own logic. Interconnected systems match up rail, barges, long-distance trucking, short-haul truck-ing, and so forth to get the item to the end user when needed. Each kind of crop and each kind of animal product travels a different path from the farm to our tables. The carbon footprint of each one varies, too.

Many farmers with smaller acreages are opting out of the agribusiness and mass production transportation system, preferring to deal as directly as they can with customers and consumers through such marketing techniques as U-pick farms, roadside markets, and community farmers markets.

American farmers are increasingly offering suburban and city folks a chance to get reconnected to the realities of rural living through agritourism. Visitors to crop and animal farms get the chance to participate in the harvest or other daily activities as part of school field trips, social events, or longer stays. Here are the most popular agritourism activities:

- Christmas tree farms
- U-pick and U-dig operations where the visitor does the harvesting
- Visits to vineyards and wineries
- Herb farms
- Pay lakes (for-fee fishing)
- Dude ranches

Each of these activities can give visitors a greater appreciation of the complexity of American agricultural practices, and a first-hand look at how farmers are trying both to lower the amount of energy they use and to use every bit of energy wisely.

The Role of Animals

The carbon impacts of raising livestock differ in many significant ways from the carbon impacts of raising plants. Growing plants is basically a three-step process that uses energy in the fields during the growing season and then uses energy again after the harvest to process the plants, then again to transport the food, feed, or fuel to its point of use. Raising animals adds more steps—the energy used to feed and care for the animals as they mature and then more energy to bring the finished animal products to the market.

 **Carbon Impact**

The advice to buy locally applies just as much to animal products as it does to fruits and vegetables. Buying locally raised chickens or buffalo or farm-raised fish saves on transportation efforts and can be substantially more energy efficient than buying meat raised hundreds of miles away, or seafood caught in an ocean thousands of miles away.

Considering these extra steps, raising livestock may appear to use excessive amounts of energy or have an extremely large carbon footprint. But there are so many different kinds of animals and animal products that generalizations about energy use in this sector often fail to take into account all factors.

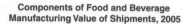

**Components of Food and Beverage
Manufacturing Value of Shipments, 2005**

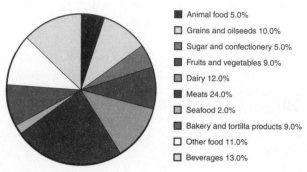

- Animal food 5.0%
- Grains and oilseeds 10.0%
- Sugar and confectionery 5.0%
- Fruits and vegetables 9.0%
- Dairy 12.0%
- Meats 24.0%
- Seafood 2.0%
- Bakery and tortilla products 9.0%
- Other food 11.0%
- Beverages 13.0%

Producing America's foods and beverages.

In many areas of the United States, agriculture and animal husbandry proceed side by side, eliminating many transportation costs and conserving energy. Thrifty farmers use the manure from animals to fertilize nearby fields, reducing the need for artificial chemical fertilizers.

Not all land forms, not all soil types, and not all climate zones are suitable for growing crops such as corn or wheat, or fruit trees, or green peppers. Raising livestock, whether it's cattle or swine or turkeys or even pond-raised catfish, may be the most appropriate, energy-efficient, and sustainable use of that land.

Many components of animal feed are leftovers or by-products of plant materials grown for some other purpose, so using these items that humans don't want to feed livestock instead prevents waste.

The search for ways to improve energy efficiency and reduce greenhouse gas emissions in the U.S. agricultural sector continues to turn up new ideas. Farming practices are changing as new ideas move from the experimental stage to practical use on the farm.

Some Farming Innovations

In their quest to use energy efficiently, U.S. farmers are trying a lot of new ideas. Here are some energy-saving innovations that are becoming common on U.S. farms:

- Minimal soil disturbance (known as conservation tillage)
- Improved irrigation methods
- GPS technology on machinery to prevent wasted efforts
- Recycled by-products and waste products
- Biofuels for farm machinery

Conservation tillage is an important way to save energy. The fewer times a farmer has to take equipment out into the field, the less fuel is consumed.

Farmers are also improving their energy efficiency by using water wisely. Sensors in the soil and satellite technology can pinpoint exactly which fields need additional moisture—and which fields don't. Other farmers are switching to crops that are more appropriate for their local weather conditions, crops that don't need any water other than what comes naturally.

def•i•ni•tion

Conservation tillage, also known as reduced tillage, means that, instead of plowing every inch of an entire field, only narrow strips just wide enough for this season's plants to grow in are plowed. Sometimes seeds can be planted directly into the soil through the covering, without plowing at all. It's a continuous cycle, with the leftovers from a previous crop or special cover crops and living mulches left in place during the off season. Then at the beginning of the new growing season, the equipment moves back and forth across the field fewer times.

GPS technology allows farmers to move equipment through fields more accurately during all times of the year. That decreases fuel usage and lowers carbon impacts.

So many innovations are being implemented, and so many new programs to encourage better energy use on America's farms are in progress, that we'll take an in-depth look at them in Chapter 9 on carbon initiatives.

A Tale of Three Pumpkins

Suppose you want a pumpkin because you and the kids are going to make a jack-o-lantern. Is it better to buy a pumpkin at your local mega grocery store today on your way home from school or work, or wait until Thursday afternoon to buy one at the farmers market in the church parking lot 2 miles away, or take a half-hour drive out to a U-pick farm on the edge of the suburbs on Saturday? This is where tracking your carbon footprint can get a bit tricky.

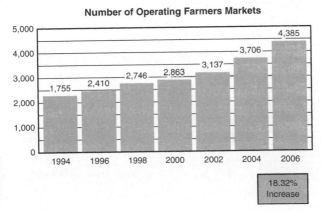

Number of Operating Farmers Markets

Year	Number
1994	1,755
1996	2,410
1998	2,746
2000	2,863
2002	3,137
2004	3,706
2006	4,385

18.32% Increase

Growth of farmers markets in the United States.

Let's take a look at the differences among the choices. The mega grocery store pumpkin may have traveled 400 or more miles to get there, but the transportation costs and carbon impact were shared among hundreds of pumpkins on that truck ride.

That's called an economy of scale and means that the carbon footprint is shared among all the pumpkins on that trip. Could be good. Then there was some packaging involved because the mega grocery stores can handle merchandise easier if it's in large containers on wooden pallets. Not so good. But you combined your commute with the shopping trip, so that's being energy efficient. Good again.

A farmers market pumpkin may have traveled only 10 miles, but it was also part of a much smaller shipment, perhaps fewer than 100 pumpkins. Fewer pumpkins, but a much shorter distance; we could be about even with the mega grocery store pumpkin. No packaging here, so that's good. You might be able to combine this shopping trip with another errand, so you're trying to be energy efficient. Could be good, too.

The U-pick pumpkin has no transportation costs and no packaging costs because it's still sitting in the field where it grew. Excellent so far—but you drove 5, 10, or 20 miles to get it. Not so good. But if you also buy some gourds, honey, and apples at the same farm, then you've made a multipurpose trip. Things are looking better. And if you take along the kids and bought a pumpkin for each one, and maybe even some extra pumpkins to take to your co-workers on Monday, you could definitely say you've made an effort at being energy efficient by buying in bulk. That's good.

In this new era of energy consciousness, shopping decisions are more complicated. And from our pumpkin example, you can see that some choices are not clear cut.

Carbon Impact

- Take along your own reusable bags on each shopping trip.
- Buy in bulk, to save trips to the source and reduce packaging.
- Reportion your bulk items into your own reusable containers.
- Cook larger recipes and then put small portions into your own reusable containers.

If you are concerned about the carbon impact of the products you find in your favorite grocery, especially the transportation that uses extra energy, let the store manager know. You can ask that signs be posted listing the origin of fresh fruits and vegetables.

Some shoppers know that peaches often come from Georgia, grapefruits come from Florida, and so forth, but it's important to put that information next to the price for everyone to see. Do you know where fresh asparagus and kiwis come from? What about bananas, radishes, and garlic?

Seasonal Adjustments

Ask any Midwesterner over the age of about 40 when strawberries are in season, and they'll tell you May and June. Older folks know that fresh corn-on-the-cob shows up in late August, fresh apples in September. Each item was a treat, something to look forward to as the seasons marched along.

Carbon Impact

Want to buy more local farm products, but don't know how to find them? Many states now have special programs to identify and promote the sale of local agricultural products, everything from honey to cheese, to cantaloupes, and even local meat products such as country ham or sausage. You can search online easily by entering the name of your state followed by the phrase "agricultural products." You should be directed to several sites that promote in-state products and offer directories of where to buy them.

But since the 1970s, Americans have grown accustomed to buying fresh fruits and vegetables in every month of the year. The following table shows the increase in American agricultural imports in billions of dollars, adjusted for inflation, using 2007 as the constant.

Value of Agricultural Imports

Year	Total Value (Billions)
1966	28.70
1976	38.52
1986	39.80
1996	43.20
2006	66.36

When you buy fresh items that are out of season in your locality, chances are, you're adding a lot to the size of your carbon footprint. Why? Because these items came from the Southern Hemisphere and traveled great distances, using fuels that emit greenhouse gases every mile of the way. You can reduce your carbon footprint with these strategies.

- Whenever possible, buy fresh foods during the season when your local crops mature and come to market.

- Adjust your menus to take advantage of fruits and vegetables as they come into season in your area.

- Check your newspaper, library, and online resources for ways to store and cook locally abundant produce when these items are in season.

Does this mean that buying canned and frozen goods out of season is carbon unfriendly? Not necessarily. Foods grown and processed within the United States still have shorter distances to travel to reach you, the customer. There are carbon impacts due to the energy used to prepare and package the foods, but because so many are processed in each batch, it can be a very efficient use of energy.

It may take less energy to process a fruit or vegetable within the United States, maintain storage for frozen or canned goods, and then transport the items slowly to local groceries over the many out-of-season months than to send one small shipment of fresh items many thousands of miles by fast overnight air freight.

The Least You Need to Know

- It takes many forms of energy to produce, process, and transport food items from the farm to your table.

- Only about 10 percent of U.S. greenhouse gas emissions come from the agricultural sector.

- American farmers are implementing many new ways to reduce their overall energy usage and to improve energy efficiency.

- Buying locally grown items can often reduce the size of your carbon footprint.

- Buying fruits and vegetables when they are in season in your area can often reduce the size of your carbon footprint.

Chapter 9

Exploring Carbon Initiatives

In This Chapter

- Understanding environmental language
- Some emissions success stories
- A world perspective on greenhouse gases
- Innovative energy technology and carbon initiatives

As more people in more places become concerned about greenhouse gases, carbon footprints, and ways to generally take better care of everything in, on, and above our planet, the push to make real changes has intensified.

Peer pressure, such as making certain activities socially unacceptable, is one approach. Another technique is developing voluntary standards and goals. Yet another method is adding the force of law.

Many of these efforts include the word *initiative* in their titles, to signify that they are the beginning of a new approach. In this chapter, we explore some of the carbon initiatives already in effect. We also take a look at a few rules and regulations being proposed. And we take a look at developing technologies that could make big differences.

A Few Bits of Energy History

Interest in the effect people have on the natural world is not new. What's different today is that we have many more ways to quickly share information among people around the globe. And in many areas, there is a new sense of urgency.

Since the 1960s, people in the United States and other countries have been trying many different ideas to help humans live satisfying lives while protecting the natural world. Some ideas have been slow in reaching general acceptance, while other ideas have already had a profound effect. Here are just a few environmental landmarks within the United States:

- 1963: Congress passes the first Clean Air Act
- 1967: Congress passes the Air Quality Act
- 1970: Congress modifies the Clean Air Act; Environmental Protection Agency (EPA) established
- 1977: U.S. Department of Energy established
- 1990: A newer Clean Air Act signed into law
- 1992: Energy Policy Act of 1992 passed
- 2005: Energy Policy Act of 2005 signed into law
- 2006: Advanced Energy Initiative passed

At the international level, the last few decades have seen many important conferences, summit meetings, and environmental initiatives centered on global climate change and greenhouse gas emissions.

Here are a few important international events:

- 1979: First World Climate Congress
- 1987: Montreal Protocol adopted (effective 1989)
- 1988: Intergovernmental Panel on Climate Change established
- 1992: Earth Summit in Rio de Janiero, Brazil
- 1997: Kyoto Protocol drafted (expires in 2012)
- 2003: European Union adoption of an Emission Trading Scheme to regulate carbon dioxide emissions (effective 2005)

Carbon Impact

The Kyoto Protocol, adopted in 1997, requires countries that sign the agreement to limit and eventually reduce their greenhouse gas limitations. The United States is not a participant. The Kyoto Protocol will expire in 2012.

Emissions trading schemes, such as the one now in effect in the European Union, are complicated financial tools and incentives. We'll take a closer look at these novel approaches in Chapter 10.

As each year goes by, government agencies at the local, state, and national levels provide more guidelines for energy use throughout the U.S. economy. Figuring out the provisions of these different laws and how they'll be implemented is far beyond the scope of this book.

Instead, we'll take a very quick look at some of the words you're likely to hear in news reports, and examine a few environmental success stories.

Carbon Concepts

You may hear people talk about a *carbon-constrained future*. In this scenario, constraints or limits in the form of rules and regulations with the force of law would control many human activities involving carbon, both at the individual level and within businesses and other organizations.

def•i•ni•tion

The phrase **carbon-constrained future** predicts that new rules and regulations will go into effect as part of a worldwide effort to reduce greenhouse gas emissions.

These anticipated carbon constraints could take the form of …

- Absolute limits on the amounts of greenhouse gases permitted in certain situations or during certain time periods.
- Financial penalties for entities that exceed the limits.
- Drastic changes in the amount of fuels or energy available from different sources.

Economic forecasters often talk about a future that includes a carbon tax as part of this carbon-constrained future, but they do not give specifics. We'll take a look at that in Chapter 10, too.

In other scenarios, forecasters envision a world in which *carbon disclosures* are required for almost everything. Corporations and government agencies—anybody and everybody—would be required to list their greenhouse gas emissions in some public way according to agreed-upon standards.

def•i•ni•tion

A **carbon disclosure** is a public statement about the amount of greenhouse gas emissions involved in an activity or enterprise. **Renewable portfolio standards** state what percentage of renewable fuels (instead of fossil fuels) must be used in a project, location, or time period.

Some laws about greenhouse gas emissions may contain *renewable portfolio standards* that specify what percentage of fossil fuels would be allowed in certain situations compared to renewable fuels. You may hear that an environmental law has a grandfather clause that gives certain entities a complete exemption or allows it more time to meet the provisions of the law. Some proposals also include requirements for mitigation, steps that will either eliminate, reduce, or control the harmful effects that activity has on the environment.

Many of these concepts have already been used successfully to address other environmental concerns.

Some Environmental Success Stories

In today's discussions of global climate change, you may run into pessimists who think the job of reducing greenhouse gases is so large there is little hope of achieving results. You'll be encouraged as you strive to lower your carbon footprint when you realize that there are already many environmental success stories.

One of those success stories involves *ozone*, a key feature of Earth's atmosphere. In the upper atmosphere, about 15 to 30 kilometers above Earth's surface, ozone absorbs some of the radiation from the sun. Ozone here is especially important because it absorbs certain harmful ultraviolet light called UVB rays. That's the good, stratospheric ozone, and it's one of the good greenhouse gases.

def•i•ni•tion

> **Ozone** is a special molecule containing three oxygen atoms. Scientists write ozone as O_3.

At Earth's surface, ozone can form from elements and compounds in the exhaust from internal combustion engines and other sources as they react with sunlight and then become trapped when certain weather conditions prevail. In high concentrations, this ozone can be harmful to animals and plants. That's the bad, tropospheric ozone.

The two kinds of ozone occur too far apart to mix. Although it might sound like a nice idea to get the ground-level ozone up to the ozone layer, there isn't any practical way to do so. But important environmental laws have already affected these two kinds of ozone.

Carbon Impact

> The United States has already reduced its production and consumption of ozone-depleting substances by 90 percent during the last 20 years.

After scientists discovered an unexpected hole in the stratospheric ozone over the South Pole, the 1987 Montreal Protocol set a timetable to phase out the use of ozone-depleting chlorofluorocarbons

(CFCs). Substitute products and entirely new technologies took their place, and the good ozone layer has begun to recover to its earlier configuration.

Scientists think it will take until the year 2050 for the good ozone layer to return to its pre-twentieth-century levels. Even though CFCs are no longer in use in many parts of the world, these man-made compounds linger for decades before they break apart. Scientists believe that dealing effectively with greenhouse gas emissions and other atmospheric effects today will also require long periods of time before substantial results can be observed.

The Clean Air Act regulates the emissions of six common pollutants, often referred to as *criteria air pollutants*. Improvements in the way exhaust and waste materials are handled in vehicles, at power plants, and in industry have dramatically reduced the amounts of many of these pollutants during the last 35 years.

def•i•ni•tion

The **criteria air pollutants** are different from greenhouse gases. The six criteria air pollutants are carbon monoxide, sulfur oxides, nitrogen oxides, lead, particulate matter, and ground-level ozone. Although ground-level ozone sounds like a greenhouse gas, it has a different role in the atmosphere than upper-level ozone, which is a good greenhouse gas.

Increased participation in recycling waste programs, antilittering campaigns, and reductions in the use of pesticides are all encouraging examples of what people are willing to do to make positive changes with far-reaching effects on the environment.

A World Perspective

Keeping track of greenhouse gas emissions is a huge endeavor. Within the United States, the Environmental Protection Agency and the Energy Information Agency gather and distribute some statistics. A voluntary group, the Climate Registry, has also begun to gather and publish information about CO_2 emissions within the United States.

Carbon Impact

In popular slang, an accountant is a bean counter. An emerging specialty in the worlds of finance and industry may be a carbon counter, someone who measures and accounts for greenhouse gas emissions. The job may end up with a more dignified name, but it could be an option for you if you like working with numbers.

But what is going on in the rest of the world? These 10 countries are the top GHG emitters in the world:

Carbon Dioxide Emissions Around the World (in Millions of Metric Tons)

	1995	2000	2005
1. USA	5,289	5,823	5,956
2. China	2,844	2,912	5,322
3. Russia	1,622	1,580	1,696
4. Japan	1,075	1,190	1,230
5. India	862	994	1,165
6. Canada	505	558	631
7. United Kingdom	555	554	577
8. South Africa	344	383	423
9. Saudi Arabia	233	289	412
10. Australia	285	352	406

Technology and Key American Initiatives

The *Hydrogen Fuel Initiative* (*HFI*) was first proposed in 2003. It later became part of official U.S. strategy with provisions included in both the Energy Policy Act of 2005 and the Advanced Energy Initiative of 2006.

def•i•ni•tion

> The **Hydrogen Fuel Initiative (HFI)** is a government program that provides funding to encourage the development and implementation of all aspects of technology needed to make using hydrogen as a fuel practical.

The aim of this initiative is to develop the technology for *fuel cells* to the point at which they can be used successfully in vehicles—and do it at a reasonable cost—by the year 2020. More than $1 billion of research money has already been earmarked to achieve this goal.

def•i•ni•tion

> In a **fuel cell,** the energy released by a reaction between a fuel such as liquid hydrogen and an oxidant, such as liquid oxygen, is converted directly and continuously into electrical energy. A fuel cell is not like a conventional battery.

Fuel cells have been around since 1839. Inventor and scientist Sir William Grove called his primitive fuel cell a gas voltaic battery. In order to make fuel cell vehicles a common transportation option, several technological breakthroughs will have to occur.

- Safety concerns must be addressed.
- A distribution network for hydrogen must be established.
- Travel range of a vehicle must meet consumer expectations (at least 300 miles between refuelings).
- Price of the fuel and the vehicle must be competitive with other choices already available.

The list is long, but in 2007, General Motors' Chevrolet division began testing a mini fleet of 100 fuel cell vehicles in real-life driving conditions. Other manufacturers are well on the way toward introducing their own versions of fuel cell vehicles.

Hydrogen fuel cells may be useful in many other situations in addition to powering vehicles. A fuel cell does not depend on fossil fuels for energy, but instead uses two readily available materials, hydrogen and oxygen. The chief advantages of fuel cells include these:

- Low or no emissions
- Quiet operation
- High efficiency

Developing many other alternative fuels with lower carbon emissions is the goal of the 25×25 Initiative. Sponsored by the Energy Future Coalition (a project of the United Nations Foundation), this campaign encourages the use of renewable energy sources such as wind, solar energy, and biofuels with the goal that these more carbon-friendly sources will make up 25 percent of the U.S. energy portfolio by the year 2025. Agricultural interests, inventors, utilities, and others across many sectors of the U.S. economy are working to achieve this goal.

Carbon Impact

Looking for more career options? If you're enjoying figuring out ways to reduce your personal carbon footprint and have a knack for science and math, you might be able to put your skills and interests to work developing new technologies to help other people. Check out some opportunities at www1.eere.energy.gov/education/careers.html.

While most of this book has concentrated on what you, as an individual, can do to reduce your carbon footprint, many businesses are also trying to modify their energy use to reduce their much larger carbon footprints.

The EPA's Climate Leaders program is a partnership between government and industry that's designed to help businesses figure out and implement workable strategies to reduce greenhouse gases. Corporate partners in this program …

- Inventory their greenhouse gases emissions.
- Report them using agreed-upon standards.
- Set goals to reduce those emissions.

- Remeasure their emissions at regular intervals.
- Report their accomplishments publicly.

You can see what these companies have pledged and how they're doing so far at www.epa.gov/stateply/index.html

Individual states and regional groups of states are also working together to develop partnerships and strategies to reduce greenhouse gas emissions.

The Least You Need to Know

- During the last 30 years, Americans and concerned people in many parts of the world have initiated a variety of efforts to reduce many kinds of harmful emissions, including greenhouse gases.

- Measuring and disclosing amounts of greenhouse gas emissions is an important step toward reducing carbon footprints in all economic sectors.

- The Hydrogen Fuel Cell Initiative and other innovative programs strive to find and implement practical substitutes for fossil fuels within the next 10 to 15 years.

- Worldwide efforts to reduce carbon footprints and the need for new technologies offer new career opportunities.

Chapter 10

Carbon Offsets, Carbon Trading, and Green Investing

In This Chapter

- How natural areas store carbon
- Carbon capture and sequestration
- Carbon credits and carbon offsets
- Carbon-conscious investing

Money is a powerful incentive. It's probably already helping motivate you to conserve energy and reduce your carbon footprint. After all, if you use less electricity and gasoline, you will save a few dollars here and there. It all adds up.

But the really big dollars are part of the international effort to reduce greenhouse gases. The financial implications of the move toward fewer carbon emissions are complex and varied. Some of these efforts will eventually affect what each of us pays for goods and services. There may also be some opportunities for you to use your money directly to help solve the problem of global climate change.

In this chapter, we follow some carbon footprint money through a few ideas and projects.

Carbon Sinks and Carbon Sequestration

Two ordinary concepts (with admittedly goofy names) are being combined with financial incentives in the struggle to reduce greenhouse gas emissions. It's a somewhat complicated path, so we start at the beginning.

Defining Terms

In Chapter 1, we took a brief look at the natural carbon cycle. We learned that carbon moves around through the rocks, soil, plants, animals, and atmosphere of our planet in a continuous cycle. Some of it moves quickly, while some of it stays in one place for millions of years.

Carbon dioxide in Earth's atmosphere is regularly absorbed into the tissues of plants, everything from giant California redwood trees to the tiniest blades of grass. Some carbon rests in soils. The world's oceans can soak up carbon from the atmosphere, and the bodies of ocean animals and plants also contain forms of carbon. When carbon molecules are not part of the atmosphere, the places they exist are called *carbon sinks*. As long as they stay underground, fossil fuels are carbon sinks.

def•i•ni•tion

A **carbon sink** (also called a carbon pool or carbon reservoir) is any area that absorbs and stores carbon from some other part of the carbon cycle. Oceans, ocean sediments, soils, and forests are carbon sinks.

In everyday usage, *sequester* means to put aside or keep separate, or to keep isolated; members of a jury are often sequestered so they can't be influenced by things that happen outside the courtroom.

In the special lingo of global climate-change scientists, the phrase *carbon sequestration* refers to ways to keep anthropogenic (man-made) carbon dioxide separate from the atmosphere.

def•i•ni•tion

Carbon sequestration is the long-term storage of carbon dioxide. Some methods are natural, such as within plant tissues, and other methods would rely on underground storage areas that involve human technology and intervention.

Some Proposals

Engineers are working on several ideas for capturing and sequestering carbon. One plan being tested now involves just the capture part. At a power plant carbon dioxide gas emissions could be captured before they entered the atmosphere and then piped over to a frozen foods processor for use as a refrigerant. Dry ice is nothing more than solidified carbon dioxide.

Many more complicated sequestration ideas propose to pump the carbon dioxide deep underground for long-term storage, perhaps in the cavities that remain after a fossil fuel has been extracted. While engineers and geologists work to solve the underground details, some very real problems above ground must be solved before these techniques are attempted. How should the surface owners be compensated for the use of area below the surface? Who owns the stored CO_2? What happens if the stored CO_2 escapes? Who is liable for any damages?

Technology and economic issues will have to move forward together before artificial sequestration becomes a workable option to reduce carbon footprints. One artificial carbon sequestration project that's almost ready to move from the drawing board to the real world is the FutureGen Initiative. Designed as a prototype electricity-generating plant, FutureGen will use coal as its fuel—but with some important differences. The FutureGen plant, which will be built in Mattoon, Illinois, will have the ability to ...

- Capture carbon dioxide before it reaches the atmosphere.
- Produce hydrogen for fuel as a by-product.
- Sequester the greenhouse gas emissions.

If FutureGen successfully demonstrates all these features, it will serve as a model to reduce carbon footprints in far-reaching ways.

Reality Checks

What about natural carbon sinks? What role will they play in reducing the accumulation of greenhouse gases in the atmosphere?

Over the years, scientists have tried several different ways to measure and then estimate how much carbon is contained in U.S. forests, soils, and other carbon sinks. The various methods produced widely differing results.

Today, due to refinements in techniques and better computing power, there is widespread agreement that all the carbon sinks in the United States (including forests, soils, and other components) absorb between one third and two thirds of a billion tons of carbon each year.

As areas of our country that were once heavily logged by clear-cutting recover and new trees planted there reach maturity, and as soil conservation practices repair damaged farmland, the potential for additional natural carbon storage increased in recent decades. Now it is decreasing again.

Carbon sinks around the world are shrinking, too, as human populations increase. Many scientists think that the amount of carbon that natural systems can absorb is near peak capacity; others believe that we've already passed that point. Scientists believe that Americans emit two to four times as much carbon as can be handled by natural carbon sinks on this continent.

 Carbon Caution

As the build-up and build-out of homes, highways, and all the parts of human society have spread out from city centers, natural areas are beginning to disappear. Our carbon sinks are shrinking.

Adding Carbon Dollars

How do carbon sinks and carbon sequestration play a role in financial incentives to reduce greenhouse gases? We'll get to that in just a short while. First we need to take a look at a few more ideas.

One way to help speed up the process of reducing greenhouse gases involves the notion of imposing caps (limits) on the amount of carbon dioxide an entity is allowed to release and then combining it with an exchange system. Participants in the system who do an extra good job of reducing their carbon emissions could trade their surplus to someone who hasn't made many reductions yet. Such a system is called *cap and trade*. In everyday terms, people refer to the allowances as permits. Over the course of many years, the permitted amounts decrease so that eventually the total of carbon emissions should decrease, too.

Going back to carbon sequestration, we need to add one more piece to the puzzle to see how it all could become part of a workable solution. In order for artificial carbon sequestration to work, the carbon dioxide must first be captured and sent to the storage area. This system goes by the names *cap and store* or capture and sequestration.

def•i•ni•tion

Two very different concepts have similar names. **Cap and trade** systems limit carbon emissions and then allow surpluses to be bought and sold to meet quotas. **Cap and store** (carbon capture and sequestration) means to capture carbon dioxide and then put it somewhere away from the atmosphere.

As if this weren't complicated enough already, we need to look at two more terms before we can put it all together. A *carbon offset* can do one of two things. It can balance a particular unit of carbon emissions with an equivalent amount of a carbon sink, or it can balance emissions in one area with savings elsewhere.

Finally, we need to know what a *carbon financial instrument (CFI) contract* is. One CFI contract equals 100 metric tons of carbon dioxide or its equivalent. The dollar value of a CFI contract varies according to market conditions. The largest market for buying and selling CFIs for all six greenhouse gases is the Chicago Climate Exchange.

def•i•ni•tion

A **carbon offset** balances greenhouse gas emissions with something that is not harmful to the atmosphere. A **carbon financial instrument (CFI) contract** puts a dollar value on a large quantity of carbon dioxide and allows entities to buy and sell them to meet emissions standards.

Now we have all the elements gathered for financial incentives to help reduce greenhouse gases. So how do they work together? And why is it important?

Well, if Company A reduces its carbon emissions far below a set limit, but Company B cannot yet do anything to reduce its carbon emissions, Company A can sell its surplus to Company B. The system is intended as a temporary fix until Company B can make its own progress.

This raises the cost of doing business for Company B—and that might be passed along to you, the consumer, when you buy Company B's goods or services. On the other hand, Company A has extra income from selling its carbon credits, so Company A might lower the price of its goods or services.

In this new economic system, when you buy from Company A, which really has a lower carbon footprint than Company B, you lower your own carbon footprint and you save money. Pretty nice, isn't it? As an individual, you might be able to get directly involved in carbon emissions trading on a much smaller scale through buying or selling a little portion of a larger CFI contract.

Just such a system is already operating in the European Union. The EU's Emission Trading Scheme (in effect since 2005) is the largest multinational cap and trade system. Industries with traditionally high carbon emissions (such as power generation and heavy manufacturing) buy and sell permitted amounts of emissions among themselves across international borders.

 Carbon Extras

> Transactions in the EU's Emission Trading Scheme are all done electronically—no paper wasted in this effort!

Here's how you can become a greenhouse gas trader:

- Buy a carbon offset when you purchase something such as an airline ticket.
- Sell a carbon offset if you own acres of forest.

Paying a person a small sum of money per acre to leave trees standing can work anywhere in the world. But the key to its success is that the forest must have a higher dollar value as a forest instead of as land for some other purpose. Owners of properly managed old-growth forests in the United

States' Appalachian region are banding together in community groups to sell the natural carbon sink capacity of their trees on the Chicago Climate Exchange. Using forests as carbon offsets or carbon sinks has to be an economically sustainable practice for the people who live nearby.

Carbon Impact

Several New England states recently formed a cap and trade system, the Regional Greenhouse Gas Initiative; to find out if you live in one of the states, visit www.rggi.org/index.htm.

Investigative reporters continue to examine the claims made by Internet companies that offer carbon offsets directly to consumers. Your money will be combined with many other people's money until there is enough to purchase a full contract. If you choose to participate, be aware that there may not be any way to verify that your money really does support a carbon sink.

Green Investing

So you've reduced your carbon footprint and saved a bunch of money by using less energy, and now you'd like to invest your money in something that will have a positive impact on the environment. Before you go any further, consider these questions:

- What can you afford to lose? Yes, lose. Be realistic—some things just don't work out, and instead of increasing in value, your investment could end up worthless.

- How long can you wait for an increase in value? Again, be realistic. If you'll be facing college costs or retiring soon, you may not be able to wait long enough for a new technology to enter the mainstream and become profitable.

- How much time on a daily, weekly, or monthly basis are you willing to spend monitoring your investment? Smart investors aren't greedy—when their investment has grown (or not grown!) to a certain point or a certain amount of time has passed, they sell and move on to something else. You must be willing to make the time to stay informed.

If you have any experience as an investor, these three concepts will be very familiar. That's because they are the questions every prudent money manager asks. You must be totally honest with yourself about the answers—and know that as the years go by, your answers will change.

Carbon Impact

Socially conscious investing—putting your dollars to use to support something you believe has more than just a monetary value—adds another layer of complexity to financial decisions.

In this new era of global markets, it is more important than ever before to seek competent advice from licensed, experienced financial professionals. Your personal carbon-friendly financial team might include your banker, your tax expert, a securities broker, a financial planner, and perhaps an attorney.

The ideas I'm going to give you here are just that—ideas to think about as you consider what to do with your money. I am not recommending any particular company or product or course of action.

Green investments can appear in just about any sector of the economy. As you look for opportunities, you might find …

- Companies with a new energy-saving technology or product.
- Companies that have a good record in reducing their own carbon footprint.
- Companies that can operate profitably even if fuel costs increase.

But beware of these green energy scams:

- If someone you don't know sends you an e-mail about a great investment opportunity, think twice before responding. Anybody with a computer can pretend to be a much bigger business than they really are.
- Watch out for high-pressure sales pitches at "free" seminars. The only reason they're free is because they make so much money selling you stuff after you walk in the door.

- Turn the other ear whenever someone offers to let you in on the "secret" of green whatever. If they're willing to tell you, it's not much of a secret, is it?

Here are some ideas about where to look for green investments.

Wind power seems likely to increase. You could ...

- Invest in an electric utility with good wind-generating capacity.
- Invest in a company that supplies blades or other wind turbine parts.

Alternative fuels look good, too. You could ...

- Invest in a railroad or trucking company involved in transporting biofuels.
- Invest in an agricultural company that supplies corn or soybean seeds.

Technology to manage energy use will be more important every year. You could ...

- Invest in a company that makes improved, multifunction electric meters.
- Invest in a company that engineers control systems or provides energy information.

If you don't have much time to investigate and research companies, you might let a mutual fund do the work for you. Many national outfits have several socially conscious funds with various "green" objectives already in operation, with proven track records. But as they say in all the legal disclaimers, past success is no guarantee of future results!

Remember, you don't have to limit your green investments to publicly traded companies.

Carbon Impact

Invent something yourself that can help save energy. Patent it, find somebody to manufacture and distribute it, and then enjoy your earnings.

You and a few savvy business partners could form your own company dedicated to some aspect of reducing greenhouse gas emissions. Every state offers business incubator programs with start-up help from seasoned business people. You might not even need to have all the money for the project yourself. Plenty of services and brokers are willing to match up venture capitalists with inventors and folks with great ideas but not a lot of cash.

The Least You Need to Know

- When technological problems are solved, capturing carbon emissions and then reusing or storing them will be a part of a wide array of efforts to reduce the impact of greenhouse gases in Earth's atmosphere.

- Financial incentives, such as cap and trade schemes and carbon offsets, may motivate people all over the world to solve the problems of greenhouse gas emissions more quickly.

- Emerging technologies and improvements in energy efficiency throughout all sectors of the world economy may offer investors opportunities to earn "green" dollars.

- When investigating opportunities, "green" investors should exercise the same (or a bit more) caution as they would when making all other financial decisions.

Chapter 11

The Five R's of Carbon-Conscious Living

In This Chapter

- How to get more involved with renewable energy
- Recycling basics and success stories
- Reuse and energy from trash
- Reducing waste everywhere
- Rethinking your choices

Throughout this book, we've looked at the principal ways Americans affect greenhouse gas emissions as we use various fuels for transportation, heating, and manufacturing. We've looked at how we generate electricity and how our use of that power in our homes, businesses, and schools contributes to greenhouse gas emissions. We've looked at energy use outdoors, energy involved in the food chain, and energy we use for entertainment.

Helpful hints to reduce your carbon footprint are scattered among these chapters to help you see how the choices you make can affect the size of your carbon footprint.

In this chapter, we look at even more ways you can use energy more sensibly by keeping the five R's in mind: renewables, recycling, reusing, reducing, and rethinking.

Financial Incentives for "Green" Electricity

One way to encourage electric utility companies to build more generating capacity that makes use of renewable resources—such as solar energy, wind, water, and biomass—is to help out with the construction costs of these innovative power plants.

def•i•ni•tion

Green tag programs offer electricity users the option to provide financial support for power generated by renewable methods.

The easiest and quickest way for you, an electricity user, to get involved in renewable energy is to participate in a *green tag* program through your local utility. You volunteer to pay an additional fee (over what your normal electricity rate is) to purchase power generated from a renewable source. Green tags are typically sold in 100 or 1,000 kilowatt-hour blocks. You can choose to buy one block or many blocks each month, depending on your budget.

Just because you sign up for a green tag program does not mean that the actual electricity arriving at your home or business really comes from a renewable energy source. Various kinds of generating plants provide electricity to the entire grid, and once that electricity enters the transmission lines, it joins all the other electricity flowing through the system—there is no way to keep it physically separate. The green power you buy may be generated 100 or more miles from where you live. Utility companies have agreed-upon standards for the reporting of green-tag purchases and renewable energy generation so that consumers can feel confident that each block of power really exists.

If a lot of customers sign up for green-tag programs, the utility can use that extra money to build even more generating capacity from renewable sources. Utilities may also trade the green tags among themselves to help finance construction projects.

Clean Renewable Energy Bonds (*CREBs*) offer utilities another way to finance large renewable energy construction projects. These bonds, authorized as part of the Energy Policy Act of 2005, differ from conventional bond issues in an important way. Instead of paying interest, the owner of the bond earns a tax credit from the federal government.

def•i•ni•tion

Clean Renewable Energy Bonds (CREBs) help finance the construction of new power plants that generate electricity from renewable sources.

 Carbon Caution

CREBs may not be appropriate for the novice investor—be sure to consult carefully with your team of financial advisors before making a decision.

Putting Solar Power to Work for You

If you want to take advantage of renewable energy right where you can see it, you can install relatively simple solar devices in your home or business. The most readily available systems are these:

- Solar water heating for household use or swimming pools
- Solar space heating
- Skylights, windows, and directional systems to spread sunlight through interior spaces

The way solar energy is captured and used in these various systems can be active or passive. In a *passive solar* system, the heat from the sun is captured and transferred directly to something else, whether it's water or air or a solid building material, or whether it's stored and released over time by physical means. In an *active solar* system, the sun's heat is captured and then transferred using mechanical means, such as fans or pumps.

def•i•ni•tion

In a **passive solar** system, no external power is needed; in an **active solar** system, mechanical devices are included.

Become a Recycler

Recycling saves more energy than you might think. Sorting your waste for the recyclable items and then buying items with recycled content can be an important step in reducing your carbon footprint.

Paper and Cardboard Recycling

Of all the stuff we throw away in a year's time, paper tops the list—it's about one third of our solid waste. Of that, we recover at least 50 percent for recycling. About half of that is in the form of corrugated boxes and other containers.

Paper cannot be recycled alone; a portion of virgin paper must be added to the recycled ingredients in order to give the finished product the proper strength. Even though making recycled paper products does require energy and other raw materials, 1 ton of recycled paper can save thousands of gallons of water, one to two dozen trees, and thousands of kilowatts of electricity. You can encourage manufacturers to continue to expand the use of recycled paper products by buying items that state on the package the percentage of recycled paper used.

*Look for these
symbols whenever
you are shopping.*

Recycling Glass

New glass is made from limestone, sand, and soda ash. Recycled glass is made from just one ingredient, old glass. Making products with recycled glass uses about 40 percent less energy than making a brand new glass product. That's because crushed old glass (called cullet) melts at a lower temperature than is needed to form fresh glass from the three ingredients. Glass doesn't wear out or weaken, so it can be used over and over again many times; there's no need to add virgin ingredients to the process as there is for paper recycling.

A little more than one fifth of the new glass produced in the United States is currently recycled. As more people become interested in the energy-saving benefits of recycling, new processes are being developed to do more with old glass than just turn it into new bottles and jars. You can find ideas and products on the Internet by searching using the phrase "recycled glass products."

Sorting glass for recycling by color is an important part of the process because once a color is added to glass, it cannot be removed; certain colors of glass provide certain advantages, so different items are packaged in either clear glass or glass of a particular color. Not all glass products can be recycled together; windowpanes, glass mirrors, light bulbs, ceramics, and some other glass products contain additional materials that make them unsuitable for common recycling processes, so they must be handled differently.

Plastics

Let's take a look at what happens when plastic is recycled. Five steps are involved:

1. Waste plastic is collected at either the curb, a center, or a business.

2. Waste plastic is sorted into similar types, and large contaminants such as metal pieces are removed.

3. The sorted waste plastic is chipped into tiny pieces.

4. The chipped plastic is washed to remove remaining contaminants such as paper labels.

5. The clean chipped plastic is melted and then formed into pellets.

These plastic pellets can be used for just about anything virgin plastic can be used for, although certain exceptions apply concerning the use of these recycled plastics with food and beverages.

The symbol for recyclable plastic is a triangle of arrows with a number from 1 to 7 inside; the number indicates what kind of plastic the item is so that similar items can be grouped together later in the recycling process.

Aluminum and Steel

Aluminum and steel are the most plentiful metals in our lives—and also among the easiest items to recycle. Instead of sending all your empties to the landfill, recycling these metal products will reduce your carbon footprint. Every day, Americans use …

- 100 million steel cans.
- 200 million aluminum beverage cans.

Recycling these two common metals saves a lot of energy. Recycling aluminum takes only 5 percent of the energy needed to make fresh aluminum the first time. Recycling steel takes a bit less than 50 percent of the energy needed to make fresh steel the first time.

Carbon Impact

The aluminum recycling system for beverage cans is so efficient that it can turn a used aluminum can back into a new one on the shelf at your local store in about eight weeks.

Sorting your household's scrap metal is an important part of becoming an active recycler. You might think of recycling metals as just something to do with food and drink cans, but it involves a lot more.

Steel is the most recycled material in the United States. Most steel cans in the United States contain 25 percent recycled steel. Steel that can be recycled comes from old cars and trucks, from old buildings, and even from old bridges, old appliances, and lots of other products. Most aluminum comes from beverage cans, but it can also come from pots and pans, lawn furniture, gutters, car parts, and many other everyday objects.

In some areas, consumers must sort the two metals, so it helps to be able to recognize them. In general, steel cans (also called tin cans because they have a thin layer of tin in them) are used for foods such as soups, tuna, green beans, pet foods, and ground coffee. Aluminum cans are used for beverages. Steel cans are attracted to magnets, but aluminum cans are not. You need only a very small magnet to do the job—one that you use to hold a photo on your refrigerator will do the job quite easily. Just bring it close to the can—aluminum will not be attracted to the magnet.

Carbon Impact

When sorting your recyclables, be careful and be honest. If you know or suspect a metal container is made of both aluminum (perhaps for the top and bottom) and steel (the sides) put it in the steel pile; steel recyclers can deal with the other metal, but aluminum recyclers can handle only aluminum.

Recycling options vary from one community to another. Some parts of the county I live in offer curbside recycling with weekly pick-ups. In my neighborhood this service is not yet available, but there is a drop-off center for paper and plastics at one of my favorite shopping districts, so I take these items along when I'll be in that area. To find out your local recycling options, start with a phone call to your local county government.

If you discover that only paper recycling is available in your area, you can "pre-cycle" by purchasing items packaged in paper instead of plastic, which isn't easily recycled in your area, as often as possible. If aluminum is easily recycled in your area, you could buy drinks in aluminum cans instead of glass bottles.

Reuse

People use a lot of words that have similar meanings in this new energy-conscious age, and it can get confusing sometimes. *Recycling* tends to refer to the kinds of things described in the previous section— a commodity that can be recovered and reprocessed into the same commodity. Used paper becomes new paper, and used steel becomes new steel.

Reuse often means making a product do a job again—often more than just one time—with little or no modification. You might reuse a paper gift bag or box several times throughout the year. You might buy cottage cheese in a plastic container and then, after you've eaten the cottage cheese, reuse the container to store paperclips.

Specialty companies are creating new products by reusing previously discarded materials. For example, old tires can be turned into these products:

- Waterproof mats
- Landscape mulches and edgings
- Drainpipes
- Barrier blocks in parking lots
- Innovative pavements

Reuse can also refer to much smaller scale projects, particularly crafts. It's relatively easy to turn an empty plastic beverage bottle into a bird feeder.

Carbon Impact

Instead of searching on the Internet for "recycled craft" ideas (you'll get more than one million answers!), include the item you want to work with, such as "recycled paper tube crafts," for a more manageable number of choices.

Reuse can involve an item that stays the same but moves to a new owner. Shopping in vintage clothing stores, antique malls, and consignment shops, and buying used items on eBay can all be part of your strategy to reduce your carbon footprint.

Trash Talk

Recycling and reuse are all ways to keep things from ending up in the trash. But what about the stuff you do throw away? The energy it takes to collect your trash and then take it somewhere else is a part of your carbon footprint you might not think about very often. Yet the average American (there he is again!) produces 4.4 pounds of waste every day. That works out to 1,606 pounds of trash during one year. Managing all that waste is a huge industry that consumes a lot of fuel—garbage trucks don't have very good fuel economy.

Preventing waste in the first place is called *source reduction*. At a manufacturing plant, careful thought to the way raw materials are purchased and handled, and how the items are packaged to send to customers can be important in reducing waste from the very beginning.

def•i•ni•tion

Source reduction means designing products and packaging so that there are fewer potential waste items from the start.

For consumers, source reduction of waste can mean making more careful purchasing decisions, such as …

- Buying in bulk to reduce packaging.
- Requesting less packaging.
- Taking along reusable shopping bags and containers.

Each of these steps means fewer things that could eventually need to be picked up by waste haulers and fewer things going to landfills.

One of the biggest success stories in redesigning a product for source reduction is the plastic soft drink bottle. Thirty years ago, a 2-liter plastic bottle weighed 68 grams. Today's 2-liter plastic soft drink bottle weighs only 51 grams. That little per bottle difference keeps 250 million pounds of plastic out of the waste stream.

Carbon Impact

Some communities are offering "pay-as-you-throw" programs to help make people more careful about what they discard. Instead of paying a flat monthly fee or annual tax for unlimited trash collection, these programs assign a waste-hauling charge per bag or other agreed-upon unit of trash during each collection cycle. The less you throw away, the less you pay.

Energy from Trash

Garbage isn't all bad—it can be put to use in two *waste-to-energy* methods that use garbage as a fuel. In a waste-to-energy facility, solid waste is burned; the heat is used to make steam for use in generating electricity or to heat buildings. By volume, garbage doesn't produce as much energy as fossil fuels do. It takes 2,000 pounds of garbage to produce the same amount of energy that 500 pounds of coal can produce. But garbage is considered a renewable resource because the stream of waste is always being replenished. About 100 waste-to-energy plants already exist in the United States, with more planned.

Burning garbage doesn't make it disappear completely. Two thousand pounds of garbage leaves behind about 500 pounds of ashy residue that must be disposed of somehow. But that is a substantially smaller amount of trash that eventually goes to a landfill.

There's another way to use garbage for energy. Landfills contribute about 25 percent of the anthropogenic methane emissions in the United States. That methane can be captured and burned to create steam to generate electricity for the grid, or for heat and energy to use in on-site manufacturing and industrial processes.

When you participate in a green tag program, you might be supporting a generating plant at a landfill.

Carbon Impact

Although individual landfill gas energy plants generate small amounts of electricity, nearly 400 of them together are already producing about 10 billion kilowatt-hours of electricity each year in the United States.

More Help and Information

You can find out more about all sorts of energy issues and get practical ideas about ways to reduce your carbon footprint from many sources. Call your local utility company and ask for information about free or low-cost energy audits for your home or business. A trained expert will come to your site and examine everything, and then make recommendations on how you can use less energy. You can match those ideas with your budget and get started saving energy right away.

Carbon Caution

Scam artists are jumping onto the energy bandwagon with bogus products and high-cost evaluations that often are inaccurate—check with your local Better Business Bureau for the latest warnings in your area.

Check with your local Cooperative Extension Service for practical information about energy use and energy conservation. County agents in every state have booklets and information sheets that are especially tailored to your local conditions. You can get connected to this network of information by entering the name of your state and the phrase "cooperative extension" in your favorite search engine.

You can find national statistics and helpful information at the Energy Information Administration's website at www.eia.doe.gov.

The Energy Kid's Page section is a good place to find basic science information. Other links on the EIA home page will connect you with energy-saving programs for homes and businesses.

An Energy Perspective

In this book, we've looked at many real-life examples of ways to reduce the size of your carbon footprint. Every day you can make progress by taking actions that …

- Reduce the amount of energy you use.
- Reduce the amount of energy and things you waste.
- Recycle and reuse products.
- Support renewable energy projects.

These physical actions are an important step. But there's another part that's just as important—rethinking what you do and how it affects the environment. It does take a bit of effort to do things differently than the way you're accustomed to doing them. And you might wonder if recycling one aluminum can today, turning off one light fixture tomorrow, and taking a vacation next month at a state park instead of flying to another country is all that important.

You might meet people who say that one action is more important than another or that one fabulous new technology will solve all problems. This book presents a lot of different ideas because the solution is more likely to come from many sources instead of just one. That's why it is important to do what you can, when you can, as often as you can.

If you can keep focused on that thought—always use energy wisely—and match your actions to your thinking, then you will be able to reduce the size of your carbon footprint. With more than 300 million people calling the United States home, what you do is vitally important. You take a few steps, your neighbor takes a few steps, and together we all accomplish something valuable.

The Least You Need to Know

- Green tag programs and renewable energy bonds encourage the construction of many kinds of power plants that emit fewer greenhouse gases.

- Metal-, plastic-, glass-, and paper-recycling programs save substantial amounts of energy and help reduce carbon footprints.

- Reducing greenhouse gas emissions throughout the United States can succeed if many individuals rethink their actions and take positive steps now.

A Brief Afterword

In this book, I've given you a lot of background information and plenty of practical tips about energy—and right about now you might be wondering, "With so many things to think about, how and where should I begin to reduce the size of my carbon footprint?"

Every day, you face plenty of opportunities to make a difference, so my final bit of advice is simple—just start somewhere! You might start this week by taking a closer look at each of your transportation choices during the next seven days. If you're already doing a good job of conserving energy during the work or school week, maybe you'll find that your weekend choices are where you can make improvements.

Many experts agree that it takes about two weeks for a new habit to become a natural part of your daily routine. Give yourself time to get accustomed to your new choice. I promise you, it *will* get easier and more familiar as the days move forward.

Once you're comfortable with your new choices, move on to another area of your life. During the following week, you could examine your food choices. You may find that with one or two different choices you could reduce the size of your carbon footprint by a big margin. In succeeding weeks, you might want to take a look at your other shopping habits, or your lighting choices, or how much water you use, and discover many other ways to make better energy choices in each of these areas of daily living.

Recently I took a close look at the things I do at home and discovered that I don't do as good a job of recycling and reusing things as I could. I came across an old saying from World War II that I put on the refrigerator door to serve as a reminder and an inspiration. Here it is:

> *Use it up*
> *Wear it out*
> *Make it do*
> *Or do without.*

That little slogan is helping me to pay closer attention to preventing waste at home and wherever I go during each week.

Remember in the introduction when I noted that every form of energy has its plusses and minuses? Now that you've looked behind the scenes in the various chapters of this book, you can better understand how trade-offs are involved in just about everything. Even something as seemingly straightforward as growing plants to produce ethanol as a substitute for fossil fuels isn't a clear-cut issue. I recently read a new study about fertilizer runoff from farm fields affecting ecosystems hundreds of miles downstream. Does that mean ethanol production from corn is a bad idea, or does it mean that better farming practices need to be discovered and put into practice right away? Should I be concerned when I fill up my car with E15?

Are you worried that, in the community where you live, recycling plastic involves so many additional transportation steps that it could have no measurable effect one way or the other on your carbon footprint? What if the products you need aren't available in glass containers that are easier to recycle locally?

Don't let such dilemmas be an excuse to do nothing. Instead, do the best you can with the information you have today. If you get better information tomorrow or a new option becomes available next month, then revise your choices accordingly.

And remember, you don't have to wait until January 1st to make a new resolution to make the very best energy choices for your situations. Any day of the year is just fine for resolving to make an effort to do something good.

Be energy wise as often as you can, and set a good example for those around you. Your choices will make a difference for all of us.

Glossary

active solar   A system to capture and distribute solar energy that includes mechanical devices.

agritourism   The term for visiting working farms and ranches. May include buying locally at a U-pick operation or at a roadside market.

annual fuel utilization efficiency (AFUE)   A rating system for heating equipment. A higher number indicates less waste.

anthropogenic emissions   The term used when human actions release elements and compounds into the atmosphere.

B2   A blend of 2 percent biodiesel and 98 percent petroleum diesel.

B20   A blend of 20 percent biodiesel and 80 percent petroleum diesel.

B100   A fuel made entirely from plant or animal products. It is a complete substitute for petroleum diesel.

biodiesel   Any diesel fuel that includes a percentage of plant or animal products.

biofuel Any fuel that includes some portion of plant or animal material.

biomass Plant materials that are used as a fuel.

British thermal unit (Btu) A standard measuring unit based on the amount of heat needed to raise the temperature of 1 pound of water by 1°F. Used to measure the output of furnaces and air conditioners.

cap and store (carbon capture and sequestration) A system that captures carbon dioxide emitted from an activity and then puts it somewhere away from the atmosphere.

cap and trade A system of rules and accounting procedures that limits carbon emissions and then allows surpluses to be bought and sold to meet quotas.

carbon-constrained future The phrase used to describe what would happen if new rules and regulations went into effect as part of a worldwide effort to reduce greenhouse gas emissions.

carbon cycle The way carbon in its many forms moves around through various parts of the natural world. Includes human contributions.

carbon dioxide (CO_2) One of the greenhouse gases. For convenience, measurements of all greenhouse gases are converted to units of carbon dioxide.

carbon disclosure A public statement about the amount of greenhouse gas emissions involved in an activity or enterprise.

carbon financial instrument (CFI) contract
A publicly traded item that puts a dollar value on a large quantity of carbon dioxide and allows entities to buy and sell these to meet emissions standards.

carbon footprint   A way to visualize the amount of greenhouse gases an activity emits.

carbon offset   An activity that balances greenhouse gas emissions with something that is less harmful to the atmosphere.

carbon sequestration   The long-term storage of carbon dioxide away from Earth's atmosphere.

carbon sink   (Also carbon pool or carbon reservoir.) Any natural area that absorbs and stores carbon from some other part of the carbon cycle.

cellulosic plant materials   Tough fibers such as stems and leaves of plants that can be used to make ethanol, a biofuel.

Clean Renewable Energy Bonds (CREBs)
A financial instrument used to help finance the construction of new power plants that generate electricity from renewable sources.

climate   Refers to the overall weather conditions in a large geographic area over long periods of time.

coal   An important fossil fuel with a high energy content that when burned emits many criteria air pollutants and greenhouse gases.

cogeneration   Refers to two kinds of dual use. One kind of cogeneration means using a waste item, such as tree bark at a paper mill, as the fuel to generate electricity. Another kind of cogeneration means to use a fuel to produce electricity and something else useful, such as steam or heat.

coke A form of coal in which most of the gases have been removed. Coke produces a particularly intense heat useful in certain manufacturing processes.

combustion The term for burning a fuel to convert its energy into heat and light. This process also releases gases from the substances contained in the fuel.

compact fluorescent light bulb (CFL) A small bulb with bent or spiraled gas-filled tubes in a single base.

conservation tillage (reduced tillage) A method of plowing and working in farm fields with the least amount of disturbance to groundcovers and soil.

criteria air pollutants A group of six emissions (carbon monoxide, sulfur oxides, nitrogen oxides, lead, particulate matter, and ground-level ozone) that have been regulated for many decades in the United States. These air pollutants are different from greenhouse gases.

distributed generation The term for small-scale power generation that is very close to the end user of that electricity.

E10 A fuel blend of 10 percent ethanol and 90 percent petroleum.

E85 A fuel blend of 85 percent ethanol and 15 percent petroleum.

electric grid   Describes the vast system of interconnected power lines that move electricity from generating plants to the eventual consumer. This term is often shortened to *grid*.

element   A substance that cannot be reduced to a simpler substance by normal chemical means.

Energy Star   A voluntary government program that certifies that an item meets high standards for energy efficiency.

envelope   The whole exterior structure of a building that separates the interior of the building from the outside elements.

ethanol   A liquid made from the starches in grains such as corn, or from the cellulose in grasses or certain kinds of trees. It is also called ethyl alcohol or grain alcohol and is useful as a fuel.

feed, food, and fuel   An important concept in agriculture used to describe the various uses for plants. Feed nourishes animals, food nourishes people, and fuel produces energy.

flex fuel vehicle (FFV)   A motor vehicle with an internal combustion engine that can be operated with either conventional petroleum fuel or blends of up to 85 percent ethanol.

fluorinated gases (CFCs and HCFCs)   Man-made combinations that affect the upper ozone layer. They include hydrofluorocarbons, perfluorocarbons, and sulfur hexafluoride.

flux The word scientists use to describe the way molecules move about, forming different combinations.

fossil fuels Organic substances (either solids, liquids, or gases) from previous geologic periods used as sources of energy. Fossil fuels are usually found underground.

fuel cell A device in which the energy released by a reaction between a fuel (such as liquid hydrogen) and an oxidant (such as liquid oxygen) is converted directly and continuously into electrical energy.

fuel consumption Describes all the fuels used for any kind of work. One vehicle may consume two kinds of fuel.

fuel economy The number of miles traveled per gallon of gas or other fuel. A common way to express this is with the term *gas mileage*.

fuel efficiency Term used to describe how much energy a fuel produces. An efficient fuel produces the most work.

generating station The place where electricity is created. Another common term is *power plant*.

geothermal energy A renewable source for heat and steam from within Earth. Heat pumps use heat near Earth's surface, while other systems make use of deeper sources, including hot springs.

global climate change Any period when Earth's atmosphere is measurably hotter or colder than in the previous period.

grandfather clause   An exemption or special provisions for entities that already exist when a new law goes into effect.

green building   An energy- and resource-efficient structure. May also refer to energy- and resource-efficient construction methods.

green roof   A specially designed roof covering that consists of plants, soil, or other growing medium, and a moisture- and root-barrier layer used to lower energy use in buildings.

green tag   The name for voluntary programs that offer electricity users a way to provide financial support for power generated by renewable methods.

greenhouse gases   A group of substances (mainly carbon dioxide, methane, nitrous oxide, and water vapor) in the atmosphere that help regulate climate and conditions at Earth's surface.

heat pump   An electric device that can provide heating in winter and cooling in summer.

high-intensity discharge (HID)   A kind of light bulb that contains gases and metals that provide very strong light very efficiently.

high-occupancy vehicle (HOV)   A passenger vehicle in which the driver is not the only passenger and mass transit vehicles such as buses.

HOV lanes (diamond lanes)   A special highway marking that restricts areas to only certain kinds of vehicles during certain times.

HVAC (heating, ventilating, and air conditioning) An abbreviation that stands for the system or systems that heat, cool, and provide fresh outdoor air to the interior of a building.

hybrid electric vehicle (HEV) A vehicle capable of switching between being propelled by electricity from a self-charging battery and using a liquid fuel in an internal combustion engine.

hydrogen fuel cell A device that converts the energy released when liquid hydrogen and liquid oxygen react into a continuous supply of electricity.

Hydrogen Fuel Initiative (HFI) A government program that provides funding to encourage the development and implementation of all aspects of technology needed to make using hydrogen as a fuel practical.

kilowatt (kW) The amount of electricity equal to 1,000 watts.

kilowatt-hour (kWh) A unit of measurement that describes the amount of electrical energy supplied by 1 kilowatt for one hour.

light-emitting diode (LED) A device that produces light from electric current passing through a tiny semiconductor instead of using gases or other extra materials.

low-pressure sodium lamp A kind of light bulb that includes gases that vaporize to create a distinctive yellowish-orange light.

LPG (liquefied petroleum gas or autogas) A fossil fuel also known as propane.

lumen   The basic unit of measurement for light.

methane (CH$_4$)   A compound that is one of the greenhouse gases. It is also the principal ingredient of natural gas.

mitigation   Any step taken to eliminate, reduce, or control the harmful effects of an activity.

molecule   The smallest physical bit of an element that can exist and still have all the characteristics of that element.

natural gas   A fossil fuel. The term can be used to describe a mixture of gases or pure methane.

nitrogen oxides (N$_2$O$_x$ or NO$_x$)   Any of a group of compounds formed when fossil fuels are burned. These gases are criteria air pollutants and play a role in the formation of smog.

nitrous oxide (N$_2$O)   A compound of two molecules of nitrogen and one molecule of oxygen. This substance is a greenhouse gas.

nonrenewable energy   Any form of energy that comes from a limited source (typically a fossil fuel) that cannot be replenished.

off-peak demand (off-peak load)   The term used to describe periods when consumers want less electricity.

organic   A term used in chemistry to describe items that contain carbon. A term used in agriculture to describe growing plants without the use of artificial fertilizers and pesticides.

organic light-emitting diode (OLED) A device that produces light as electrical energy moves through layers of materials.

ozone (O_3) A special molecule containing three oxygen atoms. At ground level, ozone is an air pollutant, but in the upper atmosphere, the ozone layer is part of the good greenhouse effect.

passive solar A system to capture and distribute solar energy in which no external power is needed.

peak demand (peak load) The term used to describe periods when consumers want the most electricity.

peat Partially decayed plant matter used as a fuel.

petroleum A fossil fuel that begins as an oily, liquid solution of hydrocarbons that can be refined into many fuels, such as gasoline, kerosene, diesel, and fuel oil.

plug-in electric vehicle (PEV) A vehicle that uses only electricity from the grid for power.

plug-in hybrid electric vehicle (PHEV) A vehicle that uses a combination of a conventional internal combustion engine with batteries that get their initial charge of electricity from the grid.

power plant The common word for an electricity-generating station.

primary carbon footprint The direct emission of greenhouse gases from an activity, such as burning diesel fuel in an engine.

R-value   A way to measure a material's ability to resist the flow of heat. Useful in comparing the effectiveness of various kinds and thicknesses of insulation.

renewable energy   Any form of energy that comes from a source that is unlimited or can easily be replenished.

renewable portfolio standards   A requirement that states what percentage of renewable fuels (instead of fossil fuels) must be used in a project, location, or time period.

seasonal energy efficiency ratio (SEER)   A rating system for central air conditioning units, in which a higher number indicates more efficient use of fuel.

secondary carbon footprint   The emissions of greenhouse gases that occur leading up to or after an activity, such as the expenditure of fuels to deliver a product to the consumer. Also known as the indirect carbon footprint.

solar heat gain coefficient (SHGC)   A number that measures a window's ability to permit solar heat gain or loss. Either a high or low SGHC is desirable, depending on a particular situation.

solid-state lighting   A way to produce light without heat as a by-product.

source reduction   The design and packaging of a product so that there are fewer potential waste items from the start.

sustainable agriculture An environmentally friendly way to farm that also brings long-term economic benefits to farm communities.

turbine A device with blades or rotors that turn at high speeds to generate electricity.

U-factor A measure of how well a window stops heat flow. A lower number is desirable.

urban heat island The tendency of buildings and other structures of a human settlement to absorb and retain more heat than the natural areas nearby.

waste-to-energy The term for electricity-generating systems and other manufacturing processes that use garbage or other wastes as a fuel.

watt (w) A basic measuring unit of electric power.

weather The word for the short-term, variable atmospheric conditions in a local area.

Xeriscaping A trademarked term for an environmentally friendly form of landscaping that uses minimal watering and requires very little maintenance.

Index

X-Y-Z